Numeracy

A Foundation Book

D1493674

Numeracy

A Foundation Book

M Steer

Pitman

PITMAN PUBLISHING LIMITED
128 Long Acre, London WC2E 9AN

A Longman Group Company

First published in Great Britain 1986

British Library Cataloguing in Publication Data
Steer, M.
   Numeracy: a foundation book.
   1. Mathematics — 1961-
   I. Title
   510   QA39.2

   ISBN 0–273–02604–6

Printed in Great Britain at The Bath Press, Avon

Contents

Preface

The methods set out in this book are the accumulated material of many years of teaching pupils in the fourth and fifth years of secondary schooling who found 'maths' difficult.

I found that the greatest obstacle to their learning was that they did not know the mechanical steps necessary to arrive at an answer; with those pupils who are 'good' at figures the logical sequences are obvious and can often be done in the head; for those having difficulty they must be 'spelt out' in the minutest detail, and this I have tried to do.

The topics covered are very basic, but once competence has been achieved then the much more interesting figure-work of vocationally orientated arithmetic can be tackled successfully. By using the appropriate commercial and industrial forms — invoices, statements, petty cash, requisitions, banking and stockroom forms, to name but a few — the use of arithmetic can become a much more practical exercise and can actually be enjoyed by those who felt they could never 'do' figures. Added to which, in the course of doing 'commercial arithmetic' the pupil will learn a great deal which will be of use to him when he leaves school or college.

The exercises in this book are in three parts — a, b and c, — and graded in difficulty. The book should be worked through using the 'a' set of exercises only; this should be followed by a complete re-run, starting with Exercise 1b; then a third time, starting with 1c. It is very easy for pupils to forget entirely the method needed for a particular arithmetic answer, and this repetitive handling of the course forms a good basis for 'stamping in' knowledge.

Some of the exercises are divided by horizontal lines. These are exercises on relatively simple topics where different types of the same method have been set out. It did not seem necessary to set a full exercise on each example and examples which could be grouped together have been collected into one exercise.

M.S.

For a free copy of the answers to *Numeracy*, *A Foundation Book*, teachers should write on College headed paper to:

Order Processing Department
Pitman Publishing Ltd
Slaidburn Crescent
SOUTHPORT PR9 9YF

1 Arithmetic

Arithmetic can be defined as 'the art of reckoning with figures'.

The figures we use are:

| 0, | 1, | 2, | 3, | 4, | 5, | 6, | 7, | 8, | 9 |

'0' is called 'nought' or 'zero'.

These figures (or digits) are used when we *count* things.

The result of counting is called a *number*.

A number is a *whole number* or *integer* — it is not a fraction, neither is it a whole number with a fraction added, such as $1\frac{1}{2}$ or 1.5.

The digits 0–9 are whole numbers, so are any combinations of these digits, for example, 91, 728, 7365, etc.

Any number consisting of, or ending with 0, 2, 4, 6, 8 is an *even* number and is divisible by 2.

Any number consisting of, or ending with 1, 3, 5, 7, 9 is an *odd* number and is *not* divisible by 2.

We can deal with numbers in four basic ways:

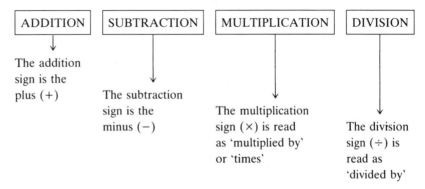

ADDITION

The addition sign is the plus (+)

SUBTRACTION

The subtraction sign is the minus (−)

MULTIPLICATION

The multiplication sign (×) is read as 'multiplied by' or 'times'

DIVISION

The division sign (÷) is read as 'divided by'

2 **Place value**

Any number greater than 9 must contain at least two figures, and the place occupied by a figure affects its value.

The 6 in 60 is worth 6×10; the 6 in 600 is worth 6×100

The values of the figures in 6 666 666 are

Millions	Hundreds of thousands	Tens of thousands	Thousands	Hundreds	Tens	Units
6 $\times$ 1 000 000	6 $\times$ 100 000	6 $\times$ 10 000	6 $\times$ 1000	6 $\times$ 100	6 $\times$ 10	6 $\times$ 1

If a number contains more than four figures it should be divided into groups of three — counting off the threes from the right and leaving a space between the groups.

Expressing numbers in words

The number 6 666 666 expressed in words becomes:

Six million, six hundred and sixty-six thousand, six hundred and sixty-six

Note that: 'hundred', 'thousand', and 'million' are written in the singular — we do not write 'six hundreds and sixty-six thousands'.

Wherever there is a space between the figures it is necessary to put in a comma *in the same place* when writing the numbers in words.

A hyphen is used in two-figure numbers which do not end in 0. 21 is written twenty-one; 43 is written forty-three; 99 is written ninety-nine.

'and' is used only when it is followed by a double figure or a single figure:

70 100 is written 'seventy thousand, one hundred'
7010   is written 'seven thousand *and ten*'
701   is written 'seven hundred *and one*'

'0' — the sign for nought or zero — is not written out in words

6 060 606 is written six million, sixty thousand, six hundred and six

6 006 600 is written six million, six thousand, six hundred

6 006 660 is written six million, six thousand, six hundred and sixty

6 600 606 is written six million, six hundred thousand, six hundred and six

---

EXERCISE 1

1a *Write out in words:*

1	18
2	25
3	20
4	53
5	67
6	101
7	219
8	300
9	613
10	2400

1b *Express in words the value of each underlined figure:*

1	1<u>8</u>
2	<u>2</u>5
3	<u>6</u>4
4	3<u>7</u>
5	8<u>1</u>3
6	1<u>9</u>2
7	9<u>1</u>41
8	<u>4</u>100
9	<u>4</u>0 096
10	17 <u>7</u>08

1c *Write out these numbers in figures:*

1	Seventy-one
2	One hundred and thirty-eight
3	Nine hundred and sixty-two
4	Eight hundred and seventeen
5	Two thousand, three hundred and ten
6	Three thousand, nine hundred and twenty-one
7	Two thousand, one hundred and four
8	Forty-one thousand, five hundred and eighteen
9	Twenty-five thousand, six hundred and twelve
10	Two hundred and thirteen thousand, two hundred

3 Whole numbers

Addition of whole numbers

The plus sign ($+$) means that the number *following* the sign must be *added to* the number preceding it.

$6 + 666$ means *add* 666 to 6

Addition questions may be written in different ways:

(a) Using the addition sign only: $6 + 666$
(b) *Add* (or *add together*) 6 and 666
(c) *Find* (or *give*) *the sum of* 6 and 666
(d) *Find* (or *give*) *the total of* 6 and 666

The Golden Rule
Write your figures neatly and do not crowd them together.

It is very important when adding and subtracting whole numbers that the figures are written out neatly under each other:

units under units
tens under tens
hundreds under hundreds
thousands under thousands,
etc.

Example $63 + 600 + 6 + 6130$

Method   1. Write figures out, remembering the golden rule

To indicate that
the figures are
to be *added* write
a plus sign
before the final
figure ——————→

63
600
6
$+ 6130$

Draw a single line
under the figures
Leave enough room for
the answer
Draw a double line under the answer

You may find it easier to write out the longest number first, followed by the other numbers in descending value order:

$$
\begin{array}{r}
6130 \\
600 \\
63 \\
+ \quad 6 \\
\hline
6799 \quad Answer \\
\hline\hline
\end{array}
$$

Check that you have written down *all* the numbers.

2. Start with the **unit** *column on the right* and add up all the figures in that column: $0 + 0 + 3 + 6 = 9$
Write the result between the lines in the units column.

3. Continue through the columns, in order, *working to the left:* the answer is 6799.

Write *Answer* after 6799.

In this example no column added up to more than 9.

In the *following* example the first three columns all add up to more than 9.

Example $66 + 666 + 6 + 6666$

Method   1. Write out the sum:

$$
\begin{array}{r}
6666 \\
666 \\
66 \\
+ \quad 6 \\
\hline
7404 \quad Answer \\
\hline\hline
122 \\
\end{array}
$$

2. Add up the units column $(6 + 6 + 6 + 6 = 24)$
Write the 4 in the units column and the 2 *under the double line in the 10s column.*

3. Add up the 10s column, *including the 2 under the line* $(6 + 6 + 6 + 2 = 20)$
Write the 0 in the 10s column and the 2 *under the double line in the 100s column.*

4. Add up the 100s column, *including the 2 under the line* $(6 + 6 + 2 = 14)$
Write the 4 in the 100s column and the 1 *under the line in the 1000s column.*

5. Add up the 1000s column, *including the 1 under the line* $(6 + 1 = 7)$ Write the 7 in the same column, thus completing the answer, which is 7404.

EXERCISE 2

2a *Add together:*
1 10 + 22
2 41 + 28
3 15 + 33
4 17 + 20
5 21 + 27
6 1 + 10 + 24
7 50 + 27 + 3
8 33 + 12 + 36
9 7 + 17 + 77 + 70
10 27 + 14 + 9 + 135

2b *Find the sum of:*
1 91, 96, and 97
2 112, 119, and 100
3 10, 700, and 60
4 106, 113, and 109
5 102, 95, and 101
6 108, 98, and 115
7 99, 104, and 110
8 111, 107, and 114
9 90, 103, and 7
10 116, 120, and 105

2c *Find the total of:*
1 4, 46, 48, and 102
2 7, 387, 80, and 117
3 56, 91, 144, and 10
4 34, 68, 18, and 96
5 1, 8, 2, and 99
6 29, 88, 27, and 92
7 15, 94, 86, and 6
8 930, 19, 48, and 9
9 93, 3, 652, and 306
10 1020, 89, 95, and 10

Subtraction of whole numbers

Subtraction is taking one quantity away from another quantity to find the difference.

The sign used is the minus $(-)$ sign which indicates that the number following the sign is to be taken away (subtracted) from the number preceding it. For example $8 - 5$ (8 take away 5) $= 3$. $8 - 5$ could also be read as '8 subtract 5'.

Subtraction questions may be written in any of the following ways:

(a) By the use of the minus sign only: $57 - 26$
(b) *Find* (or *What is*) *the difference between* 57 *and* 26?
(c) *Subtract* 26 *from* 57
(d) *How many are left if* 26 *is taken from* 57?
(e) *How much must be added to* 26 *to make* 57?
(f) *How much more than* 26 *is* 57?
(g) *How much less than* 57 *is* 26?

When subtracting involves whole numbers the numbers should be written out neatly under each other:

units under units
tens under tens
hundreds under hundreds
thousands under thousands,
etc.

and the larger number should always be written *above* the smaller number.

Example   57 − 26

Method   1. Write out the sum:
Put a minus sign   57
before the number → − 26
which is to be   ⎯⎯
31   *Answer*
subtracted   ⎯⎯

2. Starting with the units column, say '7 take away 6 = 1'
Write 1 in between the lines in the units column
Continue through the columns in order, *working to the left*

3. 31   *Answer*

This example illustrates a very easy subtraction sum

Here are three more difficult examples of subtraction

A   *Where the figure on the top line is smaller than the figure which is to be
subtracted* (Note that this applies to *individual* figures and not to the
whole number)

Example A   57 − 28

Method   1. Write out the sum: 57
− 28
⎯⎯

⎯⎯

2. Starting with the units column, say '7 take away 8' — which you
cannot do, so you must 'borrow' 10 from the 10s column. Cross off the 5
and write 4 above it, then *add 10* to the 7 in the units column which
makes it 17 (to do this write a small 1 before the 7: ¹7)

The sum now reads: 4₁
$̶5$7
− 28
⎯⎯
29   *Answer*
⎯⎯

You can now say '17 take away 8 = 9' and write the 9 in the answer
space
Continuing to the 10s column you say '4 take away 2 = 2'

29   *Answer*

B   *How to deal with a '0' which is NOT the final figure on the top line.*

Example B   507 − 28

Method   1. Write out the sum: 507
$$- \quad 28$$

2. Starting with the units column say '7 take away 8' — which you cannot do, so you must move to the 10s column where there is only a 0 — which gives you nothing to borrow from.

Now you must move left again to the 100s column and borrow from the 5; cross it off and write 4 above it.

Now move *right* to the 0; cross it off and write 9 above it.

Move *right* again and add 10 to the 7 in the units column: ^{1}7

The sum now reads:
$$
\begin{array}{r}
49_1 \\
\cancel{50}7 \\
- \ 28 \\
\hline
479 \quad \textit{Answer}
\end{array}
$$

3. It is now possible to complete the sum:

(units)   17 take away 8 = 9
(10s)   9 take away 2 = 7
(100s)   4 take away 0 = 4.

479   *Answer*

C   *How to deal with numbers ending with '0' on the top line*

Example C   5000 − 28

Method   1. Write out the sum: 5000
$$- \quad 28$$

2. Starting with the units column say '0 take away 8' — which cannot do, so you must move to the left on the top line until you reach the figure 5. Cross it off and write 4 above it—you have 'borrowed' 1.

Cross off the 0 in the 100s column and write 9 above it.

Cross off the 0 in the 10s column and write 9 above it.

Give 10 to the units column by writing a small 1 before the 0: ^{1}0

The sum now reads:

$$\begin{array}{r} 499_1 \\ \cancel{5}\cancel{0}\cancel{0}0 \\ -28 \\ \hline 4972 \quad Answer \\ \hline\hline \end{array}$$

3. It is now possible to complete the sum:

(units) $10 - 8 = 2$
(10s) $9 - 2 = 7$
(100s) $9 - 0 = 9$
(1000s) $4 - 0 = 4$

4972   *Answer*

Whenever the number on the top line ends in more than one 0, take 1 away from the first figure you come to (moving from the right), and add 1 to the 0 in the units column making it $^{1}0$ (ten).

Cross off all the 0s in between and change them to 9s:

e.g. $\begin{array}{r} 700\,000 \\ -\,156\,789 \\ \hline \\ \hline\hline \end{array}$ becomes $\begin{array}{r} 699\,99_1 \\ \cancel{7}\cancel{0}\cancel{0}\,\cancel{0}\cancel{0}0 \\ -\,156\,789 \\ \hline 543\,211 \quad Answer \\ \hline\hline \end{array}$

Note that $\begin{array}{r} 1000 \\ -\,567 \\ \hline \\ \hline\hline \end{array}$ becomes $\begin{array}{r} 099_1 \\ \cancel{1}\cancel{0}\cancel{0}0 \\ -\,567 \\ \hline 433 \quad Answer \\ \hline\hline \end{array}$

EXERCISE 3

3a	3b	3c What is left if the first number is taken from the second number?
1 59 − 36	**1** 50 − 31	**1** 55, 110
2 55 − 54	**2** 160 − 33	**2** 500, 670
3 48 − 16	**3** 400 − 101	**3** 99, 200
4 36 − 30	**4** 100 − 99	**4** 13, 20
5 97 − 86	**5** 200 − 99	**5** 139, 239
6 88 − 59	**6** 250 − 16	**6** 199, 298
7 175 − 56	**7** 1000 − 101	**7** 462, 571
8 478 − 169	**8** 6700 − 699	**8** 405, 504
9 281 − 169	**9** 4000 − 3626	**9** 1609, 1751
10 780 − 44	**10** 1400 − 1001	**10** 5000, 10 000

Multiplication of whole numbers

Multiplication is a quick way of adding up any number of quantities which all have the same value.

The sign used for multiplication ($\times$) is read as 'multiplied by' or 'times'. It means that a quantity (the *multiplicand*) has to be repeated the number of times indicated by the *multiplier*. The result is called the *product*.

For example, 18 (the multiplicand) $\times$ 3 (the multiplier) = 54 (the product).

Proof that multiplication is a form of addition is seen if instead of 18 $\times$ 3 = 54, we write down

$$
\begin{array}{r}
18 \\
18 \\
+\ 18 \\
\hline
54 \quad \textit{Answer} \\
\hline
\hline
\end{array}
$$

Sometimes multiplication questions are written in different ways:

(a) By the use of the multiplication sign only: 18 $\times$ 3
(b) *Multiply* 18 *by* 3

Short multiplication

The short multiplication method is used when the multiplier is not greater than 12.

To be able to use the short multiplication method with accuracy all the multiplication tables from 1 to 12 should be learned by heart.

If you find it really difficult to learn tables then you should master the art of preparing a multiples table. A *multiple* is a number which contains another number an exact number of times.

To prepare a multiples table, draw a square 12 cm by 12 cm in your exercise book; mark all the lines off into 1 cm lengths. Join the opposite marks and you will have 144 small squares.

Along the top, in the squares, write the numbers 1 to 12.

Down the first column on the left write the numbers 1 to 12; you will not have to write the figure 1 twice.

In the squares on the second line (going across) write the multiples of 2.

In the squares on the third line write the multiples of 3.

Proceed down the lines until you have completed the multiples of 12.

Multiples square

1	2	3	4	5	6	7	8	9	10	11	12
2	4	6	8	10	12	14	16	18	20	22	24
3	6	9	12	15	18	21	24	27	30	33	36
4	8	12	16	20	24	28	32	36	40	44	48
5	10	15	20	25	30	35	40	45	50	55	60
6	12	18	24	30	36	42	48	54	60	66	72
7	14	21	28	35	42	49	56	63	70	77	84
8	16	24	32	40	48	56	64	72	80	88	96
9	18	27	36	45	54	63	72	81	90	99	108
10	20	30	40	50	60	70	80	90	100	110	120
11	22	33	44	55	66	77	88	99	110	121	132
12	24	36	48	60	72	84	96	108	120	132	144

The multiples of 2 are

The multiples of 3 are

The multiples of 4 are

The multiples of 5 are

The multiples of 6 are

The multiples of 7 are

The multiples of 8 are

The multiples of 9 are

The multiples of 10 are

The multiples of 11 are

The multiples of 12 are

To make sure you have completed the square correctly check that the figures written along the bottom line (12, 24, 36, etc) are exactly the same as the figures written *down* the last column on the right.

If you want to know the answer to 5 × 7, put your finger on 5 in the top line and run it *down* that column until you reach the line which starts with 7 on the left; the answer will be 35.

Note that the answer to 7 × 5 is exactly the same.

Short multiplication:

Example   326 × 4

Method   1. Write out the sum:

$$\begin{array}{r} 326 \\ \times\quad 4 \\ \hline 1304 \\ \hline {\scriptstyle 1\ 2} \end{array}\quad \textit{Answer}$$

2. Start with the units on the right: 6 × 4 = 24
Write the unit 4 in the units column of the answer space.
Write the 2 under the line in the 10s column

3. Move left to the 10s column: 2 × 4 = 8

$$\begin{array}{r} +2 \text{ carried from the units} \\ \hline 10 \end{array}$$

Write the 0 in the 10s column answer space
Write 1 under the line in the 100s column

4. Move *left* to the 100s column: $3 \times 4 = 12$

$$\begin{array}{r} +1 \text{ from the 10s column} \\ \hline 13 \end{array}$$

Write 13 in the answer space

1304   *Answer*

EXERCISE 4

4a
1 18×4
2 15×5
3 16×2
4 16×6
5 18×2
6 13×3
7 14×4
8 17×5
9 14×3
10 19×4

4b
1 15×8
2 33×6
3 23×11
4 45×9
5 17×9
6 57×11
7 26×7
8 34×12
9 60×8
10 46×12

4c
1 165×3
2 711×2
3 829×4
4 673×5
5 236×7
6 148×6
7 278×8
8 392×9
9 551×12
10 484×11

Long multiplication

This method is used when the multiplier is greater than 12.

Long multiplication involves 2 numbers — when writing out the sum always write the larger number above the smaller number:

e.g. 268×23 should be written

$$\begin{array}{r} 268 \\ \times 23 \\ \hline \\ \hline\hline \end{array}$$

> Before long multiplication can be mastered you should be able to
> A   break a number down into units, 10s, 100s, 1000s, etc;
> B   multiply by 10, 100, 1000, etc;
> C   multiply by multiples of 10, 100, 1000, etc: e.g. 20, 400, 5000, etc.

A   To break a number down into units, 10s, 100s, 1000s, etc.

Example   962 breaks down into

$$\begin{array}{ll} 2 & (2 \times 1) \\ 60 & (6 \times 10) \\ 900 & (9 \times 100) \\ \hline 962 & \end{array}$$

9267 breaks down into

$$\begin{array}{ll} 7 & (7 \times 1) \\ 60 & (6 \times 10) \\ 200 & (2 \times 100) \\ 9000 & (9 \times 1000) \\ \hline 9267 \end{array}$$

B To multiply by 10, 100, 1000, etc.

Example When multiplying by 10, add a 0 to the whole number to be multiplied:
$962 \times 10 = 9620$
When multiplying by 100, add 00 to the whole number to be multiplied:
$962 \times 100 = 96\,200$
When multiplying by 1000, add 000 to the whole number to be
multiplied: $962 \times 1000 = 962\,000$

C To multiply by multiples of 10, 100, 1000, etc: e.g., 20, 400, 5000, etc.

Example *Multiplying by a multiple of 10: 962 × 20*
1. Write out the sum: 962×20
2. Cross off the 0 and transfer it to the right
3. You now have $962 \times 2\ (= 1924)$ $962 \times 2\cancel{0} = 19\,240$ *Answer*
4. Write 1924 in front of the 0

Example *Multiplying by a multiple of 100: 962 × 300*
1. Write out the sum: 962×300
2. Cross off the 00 and transfer the 00 to the right
3. You now have $962 \times 3\ (= 2886)$ $962 \times 3\cancel{00} = 288\,600$ *Answer*
4. Write 2886 in front of the 00

Example *Multiplying by a multiple of 1000: 962 × 4000*
1. Write out the sum: 962×4000
2. Cross off the 000 and transfer the 000 to the right
3. You now have $962 \times 4\ (= 3848)$ $962 \times 4\cancel{000} = 3\,848\,000$ *Answer*
4. Write 3848 in front of the 000

NB. It does not matter how many noughts there are; cross them off, count them as you do so, and transfer them to the place where your answer is to be written.

EXERCISE 5

5a   5b   5c

Break down the following numbers into units, 10s, 100s, and 1000s as shown in the example on pages 12–13

5a		5b		5c	
1	5326	**1**	1297	**1**	2952
2	1963	**2**	7654	**2**	9261
3	4291	**3**	5432	**3**	5549
4	5163	**4**	8319	**4**	3126

Multiply: *Multiply:*   *Multiply:*

5a		5b		5c	
5	61 × 10	**5**	26 × 100	**5**	508 × 10
6	61 × 20	**6**	26 × 300	**6**	75 × 100
7	61 × 30	**7**	26 × 500	**7**	18 × 1000
8	16 × 100	**8**	21 × 1000	**8**	600 × 10
9	16 × 200	**9**	21 × 2000	**9**	900 × 100
10	16 × 500	**10**	21 × 7000	**10**	50 × 1000

Example of long multiplication

Example   463 × 27

Method   1. Write down the sum ⟶

$$
\begin{array}{r}
463 \\
\times 27 \\
\hline
\end{array}
$$

2. Break down the multiplier (27): → 7 (× 463) = 3 241

→ 20 (× 463) = 9 260

27 (× 463) = 12 501   *Answer*

12 501   *Answer*

NB. With long multiplication sums it is essential that the columns of figures be kept strictly under each other — units under units, 10s under 10s, 100s under 100s, etc.

EXERCISE 6

6a   6b   6c

6a		6b		6c	
1	51 × 16	**1**	24 × 17	**1**	181 × 33
2	73 × 14	**2**	91 × 19	**2**	400 × 31
3	26 × 15	**3**	48 × 18	**3**	123 × 42
4	95 × 13	**4**	25 × 19	**4**	920 × 20
5	37 × 14	**5**	68 × 18	**5**	616 × 320
6	62 × 15	**6**	59 × 17	**6**	718 × 220
7	59 × 16	**7**	36 × 20	**7**	105 × 300
8	84 × 13	**8**	37 × 17	**8**	513 × 210
9	48 × 16	**9**	60 × 19	**9**	140 × 43
10	52 × 13	**10**	48 × 16	**10**	700 × 500

Division of whole numbers

Division is a sharing out, finding out how many times one number (the *divisor*) goes into another number (the *dividend*); the answer is called the *quotient*.

e.g.   8 (the dividend) ÷ 2 (the divisor) = 4 (the quotient)

The division sign (÷) is read as 'divided by'.
124 ÷ 4 means 124 divided into 4 parts, or 'how many times does 4 *go into* 124?'

Division questions can be asked in three different ways:
(a) By the use of the sign only: 124 ÷ 4
(b) *How many times does 4 go into 124?*
(c) *Divide 124 by 4*

Short division

The short division method is used where the divisor is not greater than 12.

> Short division presumes that you really know your multiplication tables up to the 12 times table.

Example   824 ÷ 4

Method   1. Write out the sum: 4)824

2. Deal with the first figure in the dividend:

$$\frac{206}{4)824}$$

Say '4 into 8 goes twice' (or '2 times'). There is no remainder.
Write the 2 above the line over the 8

3. Proceed to the next figure (2)
Say '4 into 2 *will not go*'.
Put a 0 above the line over the 2

4. Because the 2 could not be divided it must be combined with the following figure (4) to make 24.
Say '4 into 24 goes 6'. There is no remainder.
Write the 6 above the 4

5. 206   *Answer*

The following example presents a variation of short division — where there is a remainder 'left over' because the divisor does not divide exactly into one or more figures in the dividend, though the answer may have no remainder.

Example $632 \div 4$

Method   1. Write out the sum: $4\overline{)632}$

2. Deal with the first figure in the dividend:

$$\frac{158}{4\overline{)632}}_{2\,3}$$

Say '4 into 6 goes once, and 2 over'
Write 1 above the 6, and a small 2 before the following figure (3) making it $_2$3 (23)

3. Say '4 into 23 goes 5 and 3 over'
Write 5 above the 3 and a small 3 before the following figure (2) making it $_3$2 (32)

4. Say '4 into 32 goes 8'
Write the 8 above the 2

5. 158   *Answer*

NB. In all division sums every figure under the line (i.e. every figure in the dividend) must have a figure above it in the answer. A 0 over the first figure can be disregarded in the answer, but 0s appearing *between* figures and *at the end of the answer* must be retained.

Some division sums, such as $9 \div 3$, give an exact answer. Others leave a remainder which can be shown in either of two ways.

Example A $\dfrac{040}{5\overline{)201}}$ Remainder 1

40 Remainder 1   *Answer*

Example B $\dfrac{040}{5\overline{)201}}$ Remainder 1

In this method the remainder figure is written over the divisor to produce a fraction: $\frac{1}{5}$

$40\frac{1}{5}$ *Answer*

NB. *A remainder must always be smaller than the divisor.*

EXERCISE 7

7a
1. $12 \div 3$
2. $60 \div 2$
3. $33 \div 3$
4. $600 \div 3$
5. $568 \div 4$
6. $120 \div 4$
7. $565 \div 5$
8. $672 \div 6$
9. $497 \div 7$
10. $918 \div 9$

7b
1. $714 \div 7$
2. $786 \div 6$
3. $808 \div 8$
4. $729 \div 9$
5. $918 \div 3$
6. $1010 \div 10$
7. $156 \div 12$
8. $560 \div 10$
9. $120 \div 12$
10. $165 \div 11$

7c
Write the remainders as fractions
1. $505 \div 12$
2. $4375 \div 3$
3. $1679 \div 3$
4. $5578 \div 6$
5. $3579 \div 2$
6. $3241 \div 8$
7. $6015 \div 10$
8. $9096 \div 5$
9. $1018 \div 7$
10. $1508 \div 5$

Long division

This method is used when the divisor is greater than 12.

Before long division can be mastered you should be able to prepare a 'multiples ladder' of the divisor.

In the sum $391 \div 23$, *23 is the divisor*.

Example To prepare a multiples ladder for the divisor 23

1. Draw a vertical line with nine cross strokes

2. Write figures 1 to 9 down left side of vertical line

3. Write multiples of 23 down right side of vertical line. (This is simply done by adding 23 to every 'answer')

1	23
	+23
2	46
	+23
3	69
	+23
4	92
	+23
5	115
	+23
6	138
	+23
7	161
	+23
8	184
	+23
9	207

Example   (Long division) $391 \div 23$

Method   1. Prepare the 'multiples ladder' as above for the divisor

2. Write out the sum: $23\overline{)391}$

3.
$$
\begin{array}{r}
017 \\
23\overline{)391} \\
-23 \\
\hline
161 \\
-161 \\
\hline
000
\end{array}
$$

1×23

7×23

Say '23 into 3 will not go'. Write a 0 above the 3 over the line

4. Combine the 3 with the 9 to make 39

5. Say '23 into 39 goes once' (refer to the ladder and select the nearest multiple which is *less* than the number to be divided)
Write 1 above the 9; write -23 under the 39; take 23 from 39; the result is 16

6. Bring down the 1 from the dividend and write it after the 16

7. Say '23 into 161' — the 'ladder' tells you it goes 7 times exactly
Write 7 above the 1; write -161 under the 161; the result is NIL

8. 17   *Answer*

EXERCISE 8

8a
Prepare multiples ladders for the following numbers:

1 13
2 31
3 17
4 37
5 43
6 19
7 41
8 23
9 29
10 47

8b

Divide:

1 $169 \div 13$
2 $156 \div 13$
3 $289 \div 17$
4 $294 \div 14$
5 $209 \div 19$
6 $378 \div 18$
7 $225 \div 15$
8 $480 \div 20$
9 $480 \div 16$
10 $322 \div 14$

8c

Divide:

1 $506 \div 46$
2 $567 \div 21$
3 $1122 \div 33$
4 $810 \div 45$
5 $286 \div 22$
6 $1508 \div 58$
7 $442 \div 34$
8 $855 \div 57$
9 $1400 \div 56$
10 $1414 \div 14$

4 Fractions

Glossary

A common, proper or **vulgar fraction** is a fraction which is less than 1 in value; it is a part, or parts of a whole and is written as 2 numbers with a line between them, e.g. $\frac{1}{4}$ or 1/4 (which means 1 divided into 4 parts — or *one* quarter).

$1 \longrightarrow$ the **top** number is called the *numerator*
$4 \longrightarrow$ the **bottom** number is called the *denominator*

The **denominator** shows the number of parts into which the whole number has been divided; in $\frac{1}{4}$ the whole number has been divided into 4 parts (quarters).

The **numerator** shows the number (or quantity) of fractional parts indicated by the denominator.

$\frac{1}{4} = $ **one** quarter
$\frac{3}{4} = $ **three** quarters

Improper or **top-heavy fractions** have a numerator which is equal to or greater than the denominator.

$\frac{4}{4}$ and $\frac{5}{4}$ are improper fractions — they are not *proper* fractions because they are not *part* of a whole.

$\frac{4}{4}$ (4 quarters) is equal to **one**
$\frac{5}{4}$ (5 quarters) = 1 whole + $\frac{1}{4}$ (i.e. $1 + \frac{1}{4} = 1\frac{1}{4}$)

$1\frac{1}{4}$ is a mixed number

Mixed numbers are made up of a whole number + a fraction.
$12\frac{3}{4}$ (twelve and three-quarters) is a mixed number — it is made up of 12 whole numbers + $\frac{3}{4}$ (a fraction)

* Mixed numbers can be changed into improper fractions, and
* Improper fractions can be changed into mixed numbers.

* **Cancelling, or simplifying fractions** means the same as **reducing a fraction to its lowest terms.**

* The methods are set out on pages 21 and 22.

The value of a fraction is not altered if the numerator *and* the denominator are both divided by the same number.

For example, because 5 will divide exactly into 10 and 15, the fraction $\frac{10}{15}$ can be reduced to $\frac{2}{3}$:

$$\frac{10 \div 5}{15 \div 5} = \frac{2}{3}.$$

* **Equivalent fractions**. By cancelling down (simplifying) $\frac{10}{15}$ we can rewrite the fraction as $\frac{2}{3}$, therefore $\frac{10}{15}$ and $\frac{2}{3}$ are equivalent fractions.

Any fraction can have an equal (or equivalent) fraction if *both the numerator and denominator* are divided, or multiplied, by the same number.

$$\frac{10}{15} = \frac{10 \div 5}{15 \div 5} = \frac{2}{3} \quad \text{(using division)}$$

Similarly $\quad \dfrac{2}{3} = \dfrac{2 \times 5}{3 \times 5} = \dfrac{10}{15}$ (using multiplication)

A prime number is a number which has no factors other than itself and unity (one). The prime numbers up to 50 are:

   1, 2, 3, 5, 7, 11, 13, 17, 19, 23, 29, 31, 37, 41, 43, 47

* **A factor** is a number which will divide into another number an exact number of times.

2, 3, 4, 5, 6, 10, and 12 will all 'go into' 60 exactly and are *factors* of 60.

* **A prime factor** is a factor which is also a prime number.

The prime factors of 60 are 2, 3, and 5. Compare this with the *factors* of 60 given above.

A common factor is a number which will divide exactly into each of two or more numbers. 5 is a common factor of 20 and 35.

* **The highest common factor (HCF),** also called the **greatest common divisor (GCD),** is the greatest number which will divide exactly into two or more numbers.

A multiple is a number which contains another number an exact number of times. (See previous chapter)

A common multiple is a number which can be divided exactly by two or more numbers. 12 is a common multiple of 2, 3, 4 and 6.

* **The lowest common denominator (LCD)** or **lowest common multiple (LCM)** of two or more numbers is the *smallest* number into which they will all divide exactly. 12 is the LCD/LCM of 2, 3, 4 and 6.

* The methods for calculating these are on pages 24, 25, and 26.

Before attempting addition, subtraction, division or multiplication of fractions the following should be learned:

Changing a mixed number to an improper fraction

Example   Express $1\frac{1}{6}$ as an improper fraction

Method   1. Write down the mixed number: $1\frac{1}{6}\leftarrow$*Numerator*

2. Multiply the whole number by the denominator: $1 \times 6 = \quad 6$

Add *numerator* of fraction: $\qquad\qquad\qquad +1$

$\qquad\qquad\qquad\qquad\qquad\qquad\qquad\quad \overline{}$

$\qquad\qquad\qquad\qquad\qquad\qquad\qquad\quad 7$

$\qquad\qquad\qquad\qquad\qquad\qquad\qquad\quad \overline{\overline{}}$

3. Write the result **over** the denominator:

$\frac{7}{6}$ *Answer*

EXERCISE 9

Change the following mixed numbers to improper fractions:

9a		9b		9c	
1	$1\frac{1}{2}$	**1**	$4\frac{3}{4}$	**1**	$7\frac{1}{5}$
2	$1\frac{1}{3}$	**2**	$3\frac{1}{2}$	**2**	$8\frac{1}{10}$
3	$1\frac{2}{3}$	**3**	$5\frac{1}{6}$	**3**	$10\frac{1}{8}$
4	$2\frac{1}{2}$	**4**	$6\frac{1}{4}$	**4**	$9\frac{3}{10}$
5	$2\frac{1}{3}$	**5**	$5\frac{2}{7}$	**5**	$8\frac{1}{3}$
6	$1\frac{1}{4}$	**6**	$6\frac{1}{2}$	**6**	$9\frac{1}{3}$
7	$2\frac{2}{3}$	**7**	$7\frac{1}{3}$	**7**	$8\frac{2}{3}$
8	$1\frac{3}{4}$	**8**	$6\frac{3}{4}$	**8**	$11\frac{1}{2}$
9	$3\frac{1}{4}$	**9**	$5\frac{3}{7}$	**9**	$10\frac{3}{4}$
10	$4\frac{1}{2}$	**10**	$7\frac{1}{4}$	**10**	$9\frac{2}{3}$

Changing an improper fraction to a mixed number

Example   Express $\frac{11}{3}$ as a mixed number

Method   1. Write down the improper fraction: $\frac{11}{3}$ $\overset{\nearrow\ Numerator}{\searrow\ Denominator}$

2. Divide the numerator by the denominator: $11 \div 3 = 3$ remainder 2

The *remainder* is written **over** the denominator: $= 3 + \frac{2}{3}$

$\qquad\qquad\qquad\qquad\qquad\qquad\qquad\qquad\quad = 3\frac{2}{3}$

$3\frac{2}{3}$ *Answer*

EXERCISE 10

Change the following improper fractions to mixed numbers:

10a	10b	10c
1 $\frac{3}{2}$	**1** $\frac{5}{2}$	**1** $\frac{13}{6}$
2 $\frac{4}{3}$	**2** $\frac{7}{3}$	**2** $\frac{20}{9}$
3 $\frac{5}{3}$	**3** $\frac{9}{4}$	**3** $\frac{13}{12}$
4 $\frac{5}{4}$	**4** $\frac{11}{5}$	**4** $\frac{16}{7}$
5 $\frac{7}{4}$	**5** $\frac{13}{4}$	**5** $\frac{16}{11}$
6 $\frac{6}{5}$	**6** $\frac{9}{2}$	**6** $\frac{11}{10}$
7 $\frac{7}{5}$	**7** $\frac{8}{3}$	**7** $\frac{19}{9}$
8 $\frac{8}{5}$	**8** $\frac{11}{4}$	**8** $\frac{17}{8}$
9 $\frac{9}{5}$	**9** $\frac{11}{2}$	**9** $\frac{28}{11}$
10 $\frac{7}{6}$	**10** $\frac{10}{3}$	**10** $\frac{41}{10}$

Cancelling (down) or simplifying a fraction: reducing a fraction to its lowest (simplest) terms

Example Reduce (*or* simplify, *or* cancel down) $\frac{9}{12}$ to its lowest terms

Method 1. Write down the fraction: $\frac{9}{12}$

2. Find a common factor — here 3 divides into both 9 and 12

3. Rewrite: $\dfrac{9 \div 3}{12 \div 3} = \dfrac{3}{4}$

(NB. The divisor is not usually written out; we would simply put

$$\frac{\cancel{9}^{3}}{\cancel{12}_{4}}$$

and divide by 3 'in the head')

$\frac{3}{4}$ *Answer*

Sometimes cancelling has to be repeated more than once before the 'lowest terms' are reached.

Example Reduce $\frac{45}{60}$ to its lowest terms

Method 1. Write down the fraction: $\frac{45}{60}$

2. Divide by the common factor 3, then by the common factor 5:

$$\frac{\cancel{45}\,\cancel{15}^{3}}{\cancel{60}\,\cancel{20}_{4}} = \frac{3}{4}$$

$\frac{3}{4}$ *Answer*

EXERCISE 11

Reduce these fractions to their lowest terms:

11a		11b		11c	
1	$\frac{2}{4}$	**1**	$\frac{6}{8}$	**1**	$\frac{12}{30}$
2	$\frac{5}{10}$	**2**	$\frac{4}{10}$	**2**	$\frac{8}{24}$
3	$\frac{3}{6}$	**3**	$\frac{2}{12}$	**3**	$\frac{18}{36}$
4	$\frac{4}{8}$	**4**	$\frac{2}{14}$	**4**	$\frac{22}{33}$
5	$\frac{2}{6}$	**5**	$\frac{5}{10}$	**5**	$\frac{27}{36}$
6	$\frac{4}{6}$	**6**	$\frac{10}{12}$	**6**	$\frac{40}{50}$
7	$\frac{3}{9}$	**7**	$\frac{4}{12}$	**7**	$\frac{42}{84}$
8	$\frac{2}{8}$	**8**	$\frac{2}{10}$	**8**	$\frac{25}{30}$
9	$\frac{6}{9}$	**9**	$\frac{12}{14}$	**9**	$\frac{28}{32}$
10	$\frac{9}{12}$	**10**	$\frac{4}{14}$	**10**	$\frac{40}{60}$

Equivalent fractions

Questions on equivalent fractions involve either a missing numerator or a missing denominator.

Missing numerator

Example $\dfrac{5}{12} = \dfrac{}{36}$

Method 1. Rewrite as: $\dfrac{5 \times}{12 \times ?} = \dfrac{}{36}$

2. Work out the missing figure (?) on the **bottom** line by division: $36 \div 12 = 3$

3. Write the answer (3) in place of the ? mark, and the same figure in the space above it:

$$\dfrac{5 \times 3}{12 \times 3} = \dfrac{}{36}$$

4. Work out the missing figure on the **top** line by multiplication ($5 \times 3 = 15$). Write in the 15:

$\frac{15}{36}$ *Answer*

Missing denominator

Example $\dfrac{5}{6} = \dfrac{15}{}$

Method 1. Rewrite as: $\dfrac{5 \times ?}{6 \times} = \dfrac{15}{}$

2. Work out the missing figure (?) on the **top** line by division: $15 \div 5 = 3$

3. Write the answer (3) in place of the ? mark, and the same figure in the space below it:

$$\dfrac{5 \times 3}{6 \times 3} = \dfrac{15}{}$$

4. Work out the missing figure on the **bottom** line by multiplication ($6 \times 3 = 18$). Write in the 18:

$\frac{15}{18}$ *Answer*

EXERCISE 12

12a
Find the missing numerators:

1 $\frac{1}{2} = \frac{}{4}$
2 $\frac{1}{3} = \frac{}{6}$
3 $\frac{1}{4} = \frac{}{12}$
4 $\frac{2}{5} = \frac{}{10}$
5 $\frac{3}{4} = \frac{}{8}$
6 $\frac{4}{5} = \frac{}{15}$
7 $\frac{2}{3} = \frac{}{9}$
8 $\frac{3}{5} = \frac{}{20}$
9 $\frac{1}{6} = \frac{}{12}$
10 $\frac{3}{7} = \frac{}{14}$

12b
Find the missing numerators:

1 $\frac{1}{8} = \frac{}{24}$
2 $\frac{5}{8} = \frac{}{40}$
3 $\frac{4}{7} = \frac{}{14}$
4 $\frac{7}{8} = \frac{}{16}$
5 $\frac{6}{7} = \frac{}{28}$
6 $\frac{1}{9} = \frac{}{27}$
7 $\frac{5}{7} = \frac{}{21}$
8 $\frac{3}{8} = \frac{}{16}$
9 $\frac{7}{10} = \frac{}{30}$
10 $\frac{2}{9} = \frac{}{27}$

12c
Find the missing denominators:

1 $\frac{5}{6} = \frac{15}{}$
2 $\frac{1}{7} = \frac{2}{}$
3 $\frac{4}{9} = \frac{12}{}$
4 $\frac{1}{10} = \frac{5}{}$
5 $\frac{2}{11} = \frac{6}{}$
6 $\frac{6}{7} = \frac{24}{}$
7 $\frac{2}{7} = \frac{12}{}$
8 $\frac{5}{9} = \frac{20}{}$
9 $\frac{9}{10} = \frac{90}{}$
10 $\frac{7}{9} = \frac{21}{}$

To find the factors of a number

(NB. It is presumed at this level that factors above 12 will not be required.)

Example Find the factors of 30

Method 1. Write down all the figures from 2 to 12, with a space between:

2 3 4 5 6 7 8 9 10 11 12

2. Cross off all the numbers which will *not* divide into 30 *exactly*:

2 3 4̸ 5 6 7̸ 8̸ 9̸ 10 1̸1̸ 1̸2̸

The factors of 30 are: 2, 3, 5, 6, and 10

2, 3, 5, 6, and 10 *Answer*

To find the prime factors of a number

(NB. It is presumed at this level that prime factors above 12 will not be required.)

Example   Find the prime factors of 66

Method   1. Write down all the *prime* numbers from 2 to 12:

   2   3   5   7   11

2. Cross off all the numbers which will not divide into 66 *exactly:*

   2   3 $\cancel{5}$ $\cancel{7}$ 11

The prime factors of 66 are: 2, 3, and 11

EXERCISE 13

13a	13b	13c
Find the factors of:	Find the prime factors of:	Find the prime factors of:
1 6	**1** 6	**1** 21
2 10	**2** 10	**2** 27
3 18	**3** 18	**3** 30
4 8	**4** 8	**4** 33
5 12	**5** 12	**5** 35
6 22	**6** 22	**6** 24
7 24	**7** 24	**7** 55
8 28	**8** 28	**8** 28
9 20	**9** 20	**9** 40
10 15	**10** 15	**10** 44

To find the highest common factor (HCF) or greatest common divisor (GCD) of two or more numbers

Example   Find the HCF/GCD of 30, 45, and 60

Method   1. Write out the numbers with a space between them:

   30   45   60

2. Divide by **prime numbers** which will divide exactly into *all* the numbers. The first prime number which will do so is 3; the second is 5:

$$\begin{array}{l} \text{HCF} = \nearrow \text{ 3)}\underline{30 \qquad 45 \qquad 60} \\ 3 \times 5 \longrightarrow \text{5)}\underline{10 \qquad 15 \qquad 20} \\ \qquad\qquad\quad 2 \qquad\ 3 \qquad\ 4 \end{array}$$

At this point there is no further prime factor which will divide into all the numbers

3. The HCF/GCD is the *product* of all the prime factors which divided into all the numbers *exactly:*

$$3 \times 5 = 15$$

15 *Answer*

EXERCISE 14

14a	14b	14c
Find the HCF/GCD	Find the HCF/GCD	Find the HCF/GCD
of the following	of the following	of the following
pairs of numbers:	pairs of numbers:	sets of numbers:
1 9 and 24	**1** 8 and 24	**1** 12, 18 and 30
2 8 and 20	**2** 28 and 14	**2** 28, 42 and 70
3 12 and 18	**3** 12 and 24	**3** 16, 32 and 64
4 6 and 30	**4** 10 and 40	**4** 18, 30 and 66
5 14 and 21	**5** 18 and 30	**5** 20, 50 and 100
6 8 and 18	**6** 10 and 25	**6** 22, 66 and 88
7 9 and 36	**7** 10 and 15	**7** 24, 36 and 72
8 20 and 30	**8** 16 and 18	**8** 64, 68 and 72
9 15 and 45	**9** 6 and 33	**9** 30, 51 and 60
10 30 and 40	**10** 14 and 21	**10** 14, 56 and 112

To find the lowest common multiple (LCM) or lowest common denominator (LCD) of two or more numbers

Example Find the LCM/LCD of 2, 5, and 14

Method 1. Write out the numbers with a space between them:

2   5   14

2. Divide through by **prime** numbers:

If a number will not divide exactly, bring that number down to the next line unchanged

Prime numbers
which when
multiplied
produce the
LCM/LCD

2) 2   5   14
5) 1   5   7
7) 1   1   7
   1   1   1

3. When all the numbers are reduced to 1, write down all the prime numbers to the left of the brackets, and put multiplication signs between them: $2 \times 5 \times 7 = 70$ (This is the LCM/LCD).

70 *Answer*

EXERCISE 15

Find the Lowest Common Multiple (LCM)/Lowest Common Denominator (LCD) of the following:

15a		15b		15c	
1	2 and 3	**1**	2, 3 and 8	**1**	3, 4 and 10
2	2 and 4	**2**	2, 3 and 9	**2**	4, 5 and 6
3	2 and 5	**3**	2, 3 and 10	**3**	4, 5 and 8
4	4 and 6	**4**	2, 3 and 12	**4**	4, 5 and 10
5	3 and 5	**5**	2, 5 and 20	**5**	4, 5 and 12
6	4 and 5	**6**	3, 4 and 5	**6**	5, 6 and 12
7	3 and 12	**7**	3, 4 and 6	**7**	6, 7 and 14
8	2, 3 and 4	**8**	3, 4 and 8	**8**	6, 10 and 15
9	2, 3 and 5	**9**	3, 4 and 9	**9**	6, 12 and 18
10	2, 3 and 6	**10**	3, 4 and 12	**10**	8, 10 and 40

Addition of fractions

Remember to reduce all fractions in an answer to their lowest terms

There are four main types of addition sum relating to fractions:

A   Addition of fractions with the same denominator
B   Addition of mixed numbers with the same denominator
C   Addition of fractions with different denominators
D   Addition of mixed numbers with different denominators

A   Addition of fractions with the same denominator

(a) Where the answer is a fraction in its lowest terms

Example $\frac{1}{15} + \frac{2}{15} + \frac{8}{15}$

Method   As all these fractions have the same denominator the numerators can be added in the head $(1 + 2 + 8 = 11)$:

$$\frac{1 + 2 + 8}{15}$$ ← Add numerators only

← Denominator remains at 15

Because the addition of the numerators is *less* than the denominator, $\frac{11}{15}$ is the answer to the sum

$\frac{11}{15}$ *Answer*

(b) Where the answer can be simplified by cancelling

Example $\frac{1}{15} + \frac{4}{15} + \frac{7}{15}$

Method   1. Add the numerators as in (a) above: $1 + 4 + 7 = 12$ (i.e. $\frac{12}{15}$)

2. $\frac{\overset{4}{\cancel{12}}}{\underset{5}{\cancel{15}}}$ can be cancelled down to $\frac{4}{5}$ (divide each number by 3)

$\frac{4}{5}$ *Answer*

(c) Where the answer is an improper fraction which must be changed into a mixed number

Example $\frac{1}{5} + \frac{2}{5} + \frac{3}{5}$

Method   1. Add the numerators as in (a) above: $1 + 2 + 3 = 6$ (i.e. $\frac{6}{5}$)

2. $\frac{6}{5}$ is an improper fraction. *Change to a mixed number:* $1\frac{1}{5}$

$1\frac{1}{5}$ *Answer*

EXERCISE 16

16a

1 $\frac{1}{10} + \frac{3}{10} + \frac{5}{10}$

2 $\frac{3}{20} + \frac{7}{20} + \frac{1}{20}$

3 $\frac{2}{15} + \frac{7}{15} + \frac{2}{15}$

4 $\frac{1}{5} + \frac{2}{5} + \frac{1}{5}$

5 $\frac{1}{4} + \frac{1}{4}$

6 $\frac{1}{8} + \frac{5}{8}$

7 $\frac{2}{9} + \frac{4}{9}$

8 $\frac{1}{2} + \frac{1}{2} + \frac{1}{2}$

9 $\frac{2}{3} + \frac{2}{3}$

10 $\frac{5}{6} + \frac{5}{6} + \frac{5}{6}$

16b

1 $\frac{1}{11} + \frac{2}{11} + \frac{7}{11}$

2 $\frac{1}{12} + \frac{5}{12} + \frac{5}{12}$

3 $\frac{1}{4} + \frac{1}{4} + \frac{1}{4}$

4 $\frac{1}{16} + \frac{3}{16} + \frac{7}{16}$

5 $\frac{3}{10} + \frac{5}{10}$

6 $\frac{1}{12} + \frac{5}{12}$

7 $\frac{1}{6} + \frac{1}{6}$

8 $\frac{1}{4} + \frac{3}{4} + \frac{1}{4}$

9 $\frac{3}{5} + \frac{2}{5} + \frac{3}{5}$

10 $\frac{1}{8} + \frac{3}{8} + \frac{5}{8} + \frac{1}{8}$

16c

1 $\frac{1}{7} + \frac{3}{7} + \frac{2}{7}$

2 $\frac{1}{8} + \frac{1}{8} + \frac{5}{8}$

3 $\frac{2}{9} + \frac{4}{9} + \frac{1}{9}$

4 $\frac{1}{12} + \frac{5}{12} + \frac{1}{12}$

5 $\frac{3}{8} + \frac{1}{8}$

6 $\frac{1}{9} + \frac{5}{9}$

7 $\frac{1}{12} + \frac{5}{12}$

8 $\frac{3}{7} + \frac{2}{7} + \frac{4}{7}$

9 $\frac{1}{9} + \frac{8}{9} + \frac{2}{9}$

10 $\frac{3}{10} + \frac{7}{10} + \frac{9}{10} + \frac{3}{10}$

B   Addition of mixed numbers with the same denominator:

(a) Where the fractions do not add up to more than 1

Example $1\frac{1}{15} + \frac{3}{15} + 6\frac{1}{15}$

Method   1. Add up the whole numbers: $1 + 6 = 7$

2. Add up the fractions: $\frac{5}{15}$

3. $\frac{5}{15}$ will *cancel down* to $\frac{1}{3}$ (divide each number by 5)

4. Add whole numbers and fraction *to make one mixed number:*

$$7 + \tfrac{1}{3} = 7\tfrac{1}{3}$$

$7\tfrac{1}{3}$ *Answer*

(b) Where the fractions add up to more than 1 — i.e. to a mixed number

Example $11\tfrac{1}{5} + 1\tfrac{2}{5} + 4\tfrac{3}{5}$

Method 1. Add up the whole numbers: $11 + 1 + 4 = 16$

2. Add up the fractions: $= \tfrac{6}{5}$

3. Change $\tfrac{6}{5}$ *to a mixed number:* $(= 1\tfrac{1}{5})$

4. Add whole numbers (16) to fractions: $(1\tfrac{1}{5})$

5. $16 + 1\tfrac{1}{5} = 17\tfrac{1}{5}$

$17\tfrac{1}{5}$ *Answer*

EXERCISE 17

17a
1 $1\tfrac{1}{3} + 7\tfrac{1}{3}$
2 $3\tfrac{1}{5} + 1\tfrac{3}{5}$
3 $4\tfrac{1}{7} + 2\tfrac{2}{7}$
4 $5\tfrac{1}{9} + 1\tfrac{4}{9}$
5 $1\tfrac{1}{4} + 2\tfrac{1}{4}$

6 $1\tfrac{1}{3} + 3\tfrac{1}{3} + 2\tfrac{2}{3}$
7 $3\tfrac{2}{3} + 4\tfrac{2}{3} + 2\tfrac{1}{3}$
8 $5\tfrac{1}{2} + 2\tfrac{1}{2} + 1\tfrac{1}{2}$
9 $2\tfrac{2}{7} + 3\tfrac{3}{7} + 1\tfrac{4}{7}$
10 $1\tfrac{5}{9} + 2\tfrac{5}{9} + 1\tfrac{5}{9}$

17b
1 $1\tfrac{3}{10} + 2\tfrac{3}{10}$
2 $10\tfrac{1}{12} + 4\tfrac{5}{12}$
3 $3\tfrac{1}{14} + 9\tfrac{9}{14}$
4 $8\tfrac{2}{15} + 7\tfrac{4}{15}$
5 $5\tfrac{1}{16} + 6\tfrac{3}{16}$

6 $2\tfrac{2}{7} + 3\tfrac{3}{7} + 1\tfrac{4}{7}$
7 $1\tfrac{3}{5} + 5\tfrac{2}{5} + 2\tfrac{3}{5}$
8 $2\tfrac{1}{4} + 3\tfrac{3}{4} + 3\tfrac{1}{4}$
9 $4\tfrac{10}{11} + 6\tfrac{4}{11} + 5\tfrac{9}{11}$
10 $2\tfrac{5}{6} + 9\tfrac{5}{6} + 4\tfrac{5}{6}$

17c
1 $17\tfrac{1}{10} + 13\tfrac{7}{10} + 1\tfrac{1}{10}$
2 $14\tfrac{1}{12} + 2\tfrac{5}{12} + 3\tfrac{5}{12}$
3 $16\tfrac{2}{15} + 1\tfrac{4}{15} + 2\tfrac{4}{15}$
4 $15\tfrac{1}{16} + 3\tfrac{3}{16} + 1\tfrac{1}{16} + 2\tfrac{1}{16}$
5 $12\tfrac{1}{9} + 3\tfrac{1}{9} + 2\tfrac{1}{9}$

6 $10\tfrac{3}{11} + 2\tfrac{10}{11} + 1\tfrac{9}{11}$
7 $2\tfrac{8}{9} + 1\tfrac{1}{9} + 5\tfrac{8}{9}$
8 $10\tfrac{1}{12} + 1\tfrac{5}{12} + 2\tfrac{11}{12}$
9 $2\tfrac{1}{8} + 3\tfrac{3}{8} + 4\tfrac{5}{8}$
10 $2\tfrac{1}{12} + 3\tfrac{7}{12} + \tfrac{5}{12}$

C Addition of fractions with different denominators

Before attempting to add fractions with different denominators you must:

Find the Lowest Common Multiple (LCM)/Lowest Common Denominator (LCD) of *all* the denominators in the sum;

Find the equivalent fraction for each fraction in the sum, using the LCM/LCD you have calculated.

(a) Where the answer is a fraction in its lowest terms

Example $\tfrac{1}{2} + \tfrac{1}{6} + \tfrac{1}{8}$

Method 1. Write out sum: $\tfrac{1}{2} + \tfrac{1}{6} + \tfrac{1}{8}$

2. Find the LCM/LCD of denominators 2, 6 and 8 (24)

3. Find the equivalent fraction for each fraction, using the denominator 24:

$$\frac{1}{2} = \frac{12}{24}$$
$$\frac{1}{6} = \frac{4}{24} \qquad Add \text{ the numerators}$$
$$\frac{1}{8} = \frac{3}{24}$$

4. Add the fractions: $12 + 4 + 3 = 19$ (i.e. $\frac{19}{24}$)

$\frac{19}{24}$ *Answer*

(b) Where the answer can be simplified by cancelling

Example $\frac{1}{2} + \frac{1}{6} + \frac{1}{12}$

Method 1. Write out sum: $\frac{1}{2} + \frac{1}{6} + \frac{1}{12}$

2. Find the LCM/LCD of denominators 2, 6 and 12 (12)

3. Find the equivalent fraction for each fraction, using the denominator 12:

$$\frac{1}{2} = \frac{6}{12}; \frac{1}{6} = \frac{2}{12}; \frac{1}{12} \text{ remains unchanged}$$

4. Add the fractions: $6 + 2 + 1 = 9$ (i.e. $\frac{9}{12}$)

5. $\frac{9}{12}$ will *cancel down* to $\frac{3}{4}$ (divide each number by 3)

$\frac{3}{4}$ *Answer*

(c) Where the answer is an improper fraction which must be changed into a mixed number

$\frac{1}{2} + \frac{1}{3} + \frac{1}{4}$

1. Write out sum: $\frac{1}{2} + \frac{1}{3} + \frac{1}{4}$

2. Find the LCM/LCD of denominators 2, 3 and 4 (12)

3. Find the equivalent fraction for each fraction, using the denominator 12:

$$\frac{1}{2} = \frac{6}{12}; \frac{1}{3} = \frac{4}{12}; \frac{1}{4} = \frac{3}{12}$$

4. Add the fractions: $6 + 4 + 3 = 13$ (i.e. $\frac{13}{12}$)

5. $\frac{13}{12}$ is an improper fraction. *Change to a mixed number:* $1\frac{1}{12}$

$1\frac{1}{12}$ *Answer*

EXERCISE 18

18a	18b	18c
1 $\frac{1}{2} + \frac{1}{3}$	**1** $\frac{1}{3} + \frac{1}{12}$	**1** $\frac{3}{5} + \frac{1}{20}$
2 $\frac{1}{2} + \frac{1}{4}$	**2** $\frac{1}{4} + \frac{1}{10}$	**2** $\frac{1}{3} + \frac{1}{4}$
3 $\frac{1}{2} + \frac{2}{5}$	**3** $\frac{1}{3} + \frac{1}{5}$	**3** $\frac{2}{3} + \frac{1}{15}$
4 $\frac{1}{3} + \frac{5}{12}$	**4** $\frac{2}{7} + \frac{3}{14}$	**4** $\frac{1}{7} + \frac{4}{21}$
5 $\frac{1}{2} + \frac{3}{10}$	**5** $\frac{3}{5} + \frac{3}{20}$	**5** $\frac{3}{10} + \frac{1}{6}$
6 $\frac{1}{4} + \frac{5}{12}$	**6** $\frac{5}{6} + \frac{1}{18}$	**6** $\frac{1}{6} + \frac{1}{8} + \frac{1}{12}$
7 $\frac{3}{4} + \frac{1}{2}$	**7** $\frac{2}{3} + \frac{1}{2}$	**7** $\frac{1}{8} + \frac{5}{6} + \frac{1}{12}$
8 $\frac{1}{3} + \frac{3}{4}$	**8** $\frac{4}{5} + \frac{3}{10}$	**8** $\frac{1}{3} + \frac{1}{10} + \frac{4}{5}$
9 $\frac{5}{12} + \frac{5}{6}$	**9** $\frac{2}{3} + \frac{1}{6} + \frac{1}{2}$	**9** $\frac{2}{3} + \frac{1}{4} + \frac{1}{3}$
10 $\frac{1}{2} + \frac{1}{6} + \frac{2}{3}$	**10** $\frac{1}{2} + \frac{1}{4} + \frac{9}{20}$	**10** $\frac{1}{8} + \frac{5}{6} + \frac{5}{12}$

D Addition of mixed numbers with different denominators

(a) Where the fractions do not add up to more than 1

Example $1\frac{1}{2} + 3\frac{1}{6} + 4\frac{1}{8}$

Method 1. Write out the sum: $1\frac{1}{2} + 3\frac{1}{6} + 4\frac{1}{8}$

2. Rewrite the sum, separating whole numbers and fractions, using brackets:

$$(1 + 3 + 4) + (\tfrac{1}{2} + \tfrac{1}{6} + \tfrac{1}{8})$$

3. Deal with the fractions: find the LCM/LCD of the denominators 2, 6, and 8 (24) and find the equivalent fractions, using the denominator 24:

$$\tfrac{1}{2} = \tfrac{12}{24}; \ \tfrac{1}{6} = \tfrac{4}{24}; \ \tfrac{1}{8} = \tfrac{3}{24}$$

4. Add the fractions: $12 + 4 + 3 = 19$ (i.e. $\frac{19}{24}$)

5. Add whole numbers in bracket (see 2 above (= 8))

6. Sum now reads: $8 + \frac{19}{24} = 8\frac{19}{24}$

$8\frac{19}{24}$ *Answer*

(b) Where the fractions add up to more than 1; that is, to a mixed number

Example $1\frac{1}{2} + 3\frac{1}{3} + 2\frac{1}{4}$

Method 1. Write out the sum: $1\frac{1}{2} + 3\frac{1}{3} + 2\frac{1}{4}$

2. Rewrite the sum, separating whole numbers and fractions, using brackets:

$$(1 + 3 + 2) + (\tfrac{1}{2} + \tfrac{1}{3} + \tfrac{1}{4})$$

3. Deal with the fractions; find the LCM/LCD of the denominators 2, 3 and 4 (12), and find the equivalent fractions, using the denominator 12:

$$\tfrac{1}{2} = \tfrac{6}{12}; \ \tfrac{1}{3} = \tfrac{4}{12}; \ \tfrac{1}{4} = \tfrac{3}{12}$$

4. Add the fractions: $6 + 4 + 3 = 13$ (i.e. $\tfrac{13}{12}$)

5. Change $\tfrac{13}{12}$ *to a mixed number:* ($= 1\tfrac{1}{12}$)

6. Add whole numbers in bracket (see 2 above ($= 6$))

7. Add whole numbers (6) to fractions: ($1\tfrac{1}{12}$)

$$6 + 1\tfrac{1}{12} = 7\tfrac{1}{12}$$

$7\tfrac{1}{12}$ *Answer*

EXERCISE 19

19a		19b		19c	
1	$1\tfrac{1}{2} + 5\tfrac{1}{3}$	**1**	$1\tfrac{1}{2} + 2\tfrac{1}{5}$	**1**	$5\tfrac{1}{20} + 9\tfrac{3}{5}$
2	$2\tfrac{1}{4} + 3\tfrac{1}{2}$	**2**	$4\tfrac{1}{4} + 1\tfrac{1}{3}$	**2**	$10\tfrac{1}{3} + 6\tfrac{1}{4}$
3	$6\tfrac{2}{5} + 7\tfrac{1}{2}$	**3**	$5\tfrac{1}{5} + 3\tfrac{2}{3}$	**3**	$7\tfrac{1}{15} + 8\tfrac{2}{3}$
4	$2\tfrac{1}{2} + 6\tfrac{3}{10}$	**4**	$1\tfrac{2}{5} + 2\tfrac{1}{10}$	**4**	$4\tfrac{5}{12} + 6\tfrac{1}{4}$
5	$4\tfrac{5}{12} + 3\tfrac{1}{3}$	**5**	$2\tfrac{5}{6} + 4\tfrac{1}{18}$	**5**	$7\tfrac{3}{14} + 5\tfrac{2}{7}$
6	$2\tfrac{1}{3} + 7\tfrac{3}{4}$	**6**	$1\tfrac{10}{21} + 5\tfrac{4}{7}$	**6**	$2\tfrac{1}{8} + 3\tfrac{5}{6} + 1\tfrac{1}{12}$
7	$2\tfrac{1}{2} + 8\tfrac{3}{4}$	**7**	$6\tfrac{1}{2} + 1\tfrac{2}{3}$	**7**	$2\tfrac{1}{3} + 3\tfrac{1}{10} + 1\tfrac{4}{5}$
8	$4\tfrac{2}{3} + 6\tfrac{1}{2}$	**8**	$3\tfrac{4}{5} + 3\tfrac{3}{10}$	**8**	$4\tfrac{1}{4} + 6\tfrac{2}{3} + 1\tfrac{1}{3}$
9	$9\tfrac{1}{2} + 1\tfrac{5}{6}$	**9**	$4\tfrac{5}{6} + 2\tfrac{5}{12}$	**9**	$6\tfrac{1}{2} + 4\tfrac{7}{20} + 1\tfrac{1}{4}$
10	$5\tfrac{3}{5} + 3\tfrac{9}{10}$	**10**	$4\tfrac{1}{2} + 2\tfrac{11}{14}$	**10**	$5\tfrac{5}{12} + 5\tfrac{1}{8} + 2\tfrac{5}{6}$

Subtraction of fractions

It is useful to know that the figure 1 can be written as a fraction using any denominator required; the numerator is always the same as the denominator:

$$\tfrac{1}{1} = \tfrac{2}{2} = \tfrac{3}{3} = \tfrac{4}{4} = \tfrac{5}{5} = \tfrac{6}{6} = \tfrac{7}{7} = \tfrac{8}{8} = \tfrac{9}{9} \text{ etc.}$$

This makes the subtraction of any fraction from a whole number much simpler:

Example $1 - \tfrac{8}{9}$

Method   Using the denominator of the *fraction*, rewrite as:

$$\tfrac{9}{9} - \tfrac{8}{9} = \tfrac{1}{9} \quad \textit{Answer}$$

There are five main types of subtraction sum relating to fractions:

A   Subtraction of one fraction from another fraction, both having the same denominator

B   Subtraction of a fraction, or a mixed number, from a mixed number, both having the same denominator

C   Subtraction of one fraction from another fraction, not having the same denominator

D   Subtraction of a fraction, or a mixed number, from a mixed number not having the same denominator

E   Subtraction of a fraction from a whole number greater than 1

A   Subtraction of one fraction from another fraction, both having the same denominator

(a) Where the answer is a fraction in its lowest terms

Example $\frac{5}{7} - \frac{3}{7}$

Method   With the same denominator there is no problem:

$$\frac{5-3}{7} = \frac{2}{7} \quad Answer$$

(b) Where the answer can be simplified by cancelling

Example $\frac{5}{6} - \frac{1}{6}$

Method $\dfrac{5-1}{6} = \dfrac{4}{6}$ (*Cancel down* — divide each number by 2) = $\frac{2}{3}$

$\frac{2}{3}$ *Answer*

B   Subtraction of a fraction, or a mixed number, from a mixed number, both having the same denominator

(a) Where the fraction to be deducted is the smaller of the two fractions

Example $2\frac{3}{4} - 1\frac{1}{4}$

Method   1. Write out the sum: $2\frac{3}{4} - 1\frac{1}{4}$

2. Rewrite the sum, separating whole numbers and fractions, using brackets:

$$(2 - 1) + (\tfrac{3}{4} - \tfrac{1}{4})$$

$$= \quad 1 \quad + \quad \tfrac{2}{4}$$

> Because you will have to add the results of the two brackets, put a *plus* sign between

3. *Cancel down fraction (divide by 2)* $= \tfrac{1}{2}$

4. Sum now reads: $1 + \tfrac{1}{2} = 1\tfrac{1}{2}$

$1\tfrac{1}{2}$ *Answer*

(b) Where the fraction to be deducted is the larger of the two fractions

Example $5\tfrac{1}{7} - 1\tfrac{5}{7}$

Method   1. Write out the sum: $5\tfrac{1}{7} - 1\tfrac{5}{7}$

2. Rewrite the sum, separating whole numbers and fractions, using brackets with a *plus* sign between:

$$(5 - 1) + (\tfrac{1}{7} - \tfrac{5}{7})$$

Because $\tfrac{5}{7}$ is greater than $\tfrac{1}{7}$ it is not possible to deduct it — **one** must be borrowed from the whole numbers and 'given' to the fractions; reduce the first figure in the 'whole numbers' bracket by 1, and give 1 to the first fraction inside the second bracket:

$$(\cancelto{4}{5} - 1) + (1\tfrac{1}{7} - \tfrac{5}{7})$$

3. Change $1\tfrac{1}{7}$ to an improper fraction: $= \tfrac{8}{7}$

4. Rewrite sum: $(4 - 1) + (\tfrac{8}{7} - \tfrac{5}{7})$

$$= 3 + \tfrac{3}{7} = 3\tfrac{3}{7}$$

$3\tfrac{3}{7}$ *Answer*

EXERCISE 20

20a

1 $\tfrac{3}{5} - \tfrac{1}{5}$
2 $\tfrac{6}{7} - \tfrac{2}{7}$

3 $\tfrac{3}{4} - \tfrac{1}{4}$
4 $\tfrac{11}{12} - \tfrac{1}{12}$

5 $6\tfrac{2}{3} - 4\tfrac{1}{3}$
6 $5\tfrac{3}{10} - 3\tfrac{1}{10}$

20b

1 $\tfrac{5}{9} - \tfrac{3}{9}$
2 $\tfrac{9}{11} - \tfrac{3}{11}$

3 $\tfrac{5}{6} - \tfrac{1}{6}$
4 $\tfrac{7}{12} - \tfrac{1}{12}$

5 $8\tfrac{5}{7} - 2\tfrac{4}{7}$
6 $9\tfrac{5}{6} - 5\tfrac{1}{6}$

20c

1 $\tfrac{10}{11} - \tfrac{1}{11}$
2 $\tfrac{7}{9} - \tfrac{2}{9}$

3 $\tfrac{7}{8} - \tfrac{3}{8}$
4 $\tfrac{5}{12} - \tfrac{1}{12}$

5 $10\tfrac{7}{9} - 1\tfrac{5}{9}$
6 $7\tfrac{11}{12} - 4\tfrac{1}{12}$

EXERCISE 20 (*continued*)

20a	20b	20c
7 $4\frac{1}{3} - 2\frac{2}{3}$	**7** $5\frac{5}{7} - 2\frac{6}{7}$	**7** $4\frac{8}{11} - 2\frac{9}{11}$
8 $4\frac{2}{5} - 2\frac{3}{5}$	**8** $5\frac{5}{9} - 3\frac{7}{9}$	**8** $6\frac{1}{5} - 4\frac{4}{5}$
9 $7\frac{1}{12} - 5\frac{5}{12}$	**9** $7\frac{1}{10} - 4\frac{9}{10}$	**9** $6\frac{3}{10} - 3\frac{7}{10}$
10 $7\frac{1}{4} - 5\frac{3}{4}$	**10** $7\frac{1}{6} - 5\frac{5}{6}$	**10** $3\frac{1}{4} - 2\frac{3}{4}$

C   Subtraction of one fraction from another fraction, not having the same denominator

(a) Where the answer is a fraction in its lowest terms

Example $\frac{3}{8} - \frac{1}{3}$

Method   1. Write out the sum: $\frac{3}{8} - \frac{1}{3}$

2. Find the LCM/LCD of the denominators 8 and 3 (24)

3. Find the equivalent fraction for each fraction, using the denominator 24:

$$\frac{3}{8} = \frac{9}{24}$$

$$\frac{1}{3} = \frac{8}{24}$$

4. Sum now reads: $\frac{9}{24} - \frac{8}{24} = \frac{1}{24}$

$\frac{1}{24}$ *Answer*

(b) Where the answer can be simplified by cancelling

Example $\frac{9}{14} - \frac{1}{7}$

Method   1. Write out the sum: $\frac{9}{14} - \frac{1}{7}$

2. Find the LCM/LCD of the denominators 14 and 7 (14)

3. Find the equivalent fraction for each fraction, using the denominator 14:

$\frac{9}{14}$ remains unchanged; $\frac{1}{7} = \frac{2}{14}$

4. Sum now reads: $\frac{9}{14} - \frac{2}{14} = \frac{7}{14}$

5. *Cancel down* (divide each number by 7) $= \frac{1}{2}$

$\frac{1}{2}$ *Answer*

EXERCISE 21

21a

1. $\frac{1}{2} - \frac{1}{3}$
2. $\frac{1}{2} - \frac{1}{8}$
3. $\frac{1}{2} - \frac{1}{5}$
4. $\frac{1}{2} - \frac{2}{7}$
5. $\frac{1}{3} - \frac{1}{5}$

6. $\frac{1}{2} - \frac{1}{4}$
7. $\frac{2}{3} - \frac{1}{6}$
8. $\frac{1}{6} - \frac{1}{18}$
9. $\frac{3}{4} - \frac{3}{20}$
10. $\frac{3}{4} - \frac{1}{12}$

21b

1. $\frac{1}{3} - \frac{1}{7}$
2. $\frac{3}{4} - \frac{1}{3}$
3. $\frac{3}{4} - \frac{2}{3}$
4. $\frac{5}{8} - \frac{1}{2}$
5. $\frac{7}{8} - \frac{1}{4}$

6. $\frac{11}{12} - \frac{1}{6}$
7. $\frac{7}{12} - \frac{1}{3}$
8. $\frac{9}{10} - \frac{2}{5}$
9. $\frac{11}{12} - \frac{2}{3}$
10. $\frac{5}{6} - \frac{1}{3}$

21c

1. $\frac{5}{6} - \frac{4}{5}$
2. $\frac{6}{7} - \frac{1}{2}$
3. $\frac{1}{2} - \frac{1}{7}$
4. $\frac{8}{9} - \frac{1}{2}$
5. $\frac{1}{2} - \frac{2}{9}$

6. $\frac{3}{5} - \frac{4}{15}$
7. $\frac{13}{14} - \frac{3}{7}$
8. $\frac{1}{6} - \frac{1}{10}$
9. $\frac{3}{10} - \frac{1}{6}$
10. $\frac{5}{12} - \frac{1}{4}$

D   Subtraction of a fraction, or a mixed number, from a mixed number not having the same denominator

(a) Where the fraction to be deducted is the smaller of the two fractions

Example $5\frac{5}{8} - 2\frac{1}{2}$

Method   1. Write out the sum: $5\frac{5}{8} - 2\frac{1}{2}$

2. Rewrite the sum, separating whole numbers and fractions, using brackets with a *plus* sign between:

$(5 - 2) + (\frac{5}{8} - \frac{1}{2})$

3. Deal with the fractions; find the LCM/LCD of the denominators 8 and 2 (8) and find the equivalent fractions, using the denominator 8:

$\frac{5}{8}$ remains unchanged; $\frac{1}{2} = \frac{4}{8}$

4. Sum now reads: $(5 - 2) + (\frac{5}{8} - \frac{4}{8})$

$$= \quad 3 \quad + \quad \frac{1}{8} \quad = 3\frac{1}{8}$$

$3\frac{1}{8}$ *Answer*

EXERCISE 22

22a

1. $2\frac{1}{2} - 1\frac{1}{3}$
2. $3\frac{1}{2} - 1\frac{1}{8}$
3. $4\frac{3}{5} - 2\frac{1}{2}$

22b

1. $3\frac{5}{6} - 1\frac{1}{5}$
2. $2\frac{1}{4} - 1\frac{1}{5}$
3. $4\frac{2}{3} - 1\frac{1}{4}$

22c

1. $3\frac{2}{3} - 1\frac{1}{2}$
2. $2\frac{7}{8} - 1\frac{1}{12}$
3. $5\frac{3}{4} - 2\frac{3}{5}$

EXERCISE 22 (*continued*)

22a

4 $5\frac{1}{4} - \frac{1}{5}$
5 $6\frac{3}{4} - \frac{1}{3}$
6 $5\frac{7}{12} - 1\frac{1}{3}$
7 $6\frac{2}{3} - 2\frac{1}{6}$
8 $2\frac{1}{6} - 1\frac{1}{18}$
9 $5\frac{7}{12} - 2\frac{1}{4}$
10 $3\frac{7}{20} - 1\frac{1}{4}$

22b

4 $5\frac{1}{2} - \frac{1}{3}$
5 $4\frac{3}{4} - \frac{3}{5}$
6 $3\frac{8}{9} - 1\frac{7}{18}$
7 $2\frac{9}{10} - 1\frac{3}{20}$
8 $2\frac{3}{5} - 1\frac{4}{15}$
9 $2\frac{7}{12} - 1\frac{1}{3}$
10 $10\frac{5}{6} - 1\frac{1}{3}$

22c

4 $6\frac{4}{5} - \frac{3}{4}$
5 $2\frac{1}{2} - \frac{1}{4}$
6 $6\frac{2}{3} - 4\frac{1}{15}$
7 $4\frac{11}{12} - 1\frac{3}{4}$
8 $9\frac{17}{18} - 1\frac{1}{9}$
9 $3\frac{9}{14} - 2\frac{1}{7}$
10 $11\frac{23}{24} - 6\frac{1}{12}$

(b) Where the fraction to be deducted is the larger of the two fractions

Example $9\frac{1}{3} - 3\frac{5}{8}$

Method 1. Write out the sum: $9\frac{1}{3} - 3\frac{5}{8}$

2. Rewrite the sum, separating whole numbers and fractions, using brackets with a *plus* sign between:

$(9 - 3) + (\frac{1}{3} - \frac{5}{8})$

3. Find the LCM/LCD of the denominators 3 and 8 (24)

4. Find the equivalent fractions for each fraction, using the denominator 24:

$\frac{1}{3} = \frac{8}{24}; \frac{5}{8} = \frac{15}{24}$

5. Sum now reads: $(9 - 3) + (\frac{8}{24} - \frac{15}{24})$

6. Because $\frac{15}{24}$ is greater than $\frac{8}{24}$ it is not possible to deduct it — **one** must be borrowed from the whole numbers and 'given' to the fractions; reduce the first figure in the 'whole numbers' bracket by 1, and give 1 to the first fraction inside the second bracket:

7. $(\cancel{9}^{8} - 3) + (1\frac{8}{24} - \frac{15}{24})$

8. Change $1\frac{8}{24}$ to an improper fraction: $= \frac{32}{24}$

9. Rewrite sum: $(8 - 3) + (\frac{32}{24} - \frac{15}{24})$

10. $= 5 + \frac{17}{24} = 5\frac{17}{24}$

$5\frac{17}{24}$ *Answer*

E Subtraction of a fraction or a mixed number from a whole number greater than 1

(a) Subtraction of a fraction from a whole number greater than 1

Example $178 - \frac{5}{6}$

Method   1. Rewrite sum: $(178 - 1) + (1 - \frac{5}{6})$

2. Rewrite whole number in second bracket as a fraction, using denominator of the fraction:

$$= (178 - 1) + (\tfrac{6}{6} - \tfrac{5}{6})$$

$$= \quad 177 \quad + \quad \tfrac{1}{6} \quad = 177\tfrac{1}{6}$$

$177\tfrac{1}{6}$ *Answer*

(b) Subtraction of a mixed number from a whole number greater than 1

Example $56 - 3\tfrac{3}{4}$

Method   1. Rewrite the sum: $(56 - 3) - \tfrac{3}{4}$

$$= \quad 53 \quad - \tfrac{3}{4}$$

2. 'Borrow' 1 from the whole number and 'give' it to the fraction:

$(53 - 1) + (1 - \tfrac{3}{4})$ (Note the *plus* sign between the brackets)

3. Rewrite whole number in second bracket as a fraction, using denominator of the fraction:

$$= (53 - 1) + (\tfrac{4}{4} - \tfrac{3}{4}) = 52 + \tfrac{1}{4} = 52\tfrac{1}{4}$$

$52\tfrac{1}{4}$ *Answer*

EXERCISE 23

23a	23b	23c
1 $7\tfrac{1}{4} - 4\tfrac{1}{2}$	**1** $5\tfrac{1}{4} - 2\tfrac{2}{3}$	**1** $6\tfrac{1}{6} - 1\tfrac{2}{3}$
2 $6\tfrac{1}{2} - 2\tfrac{3}{4}$	**2** $4\tfrac{3}{5} - 2\tfrac{3}{4}$	**2** $12\tfrac{1}{18} - 5\tfrac{1}{6}$
3 $4\tfrac{3}{5} - 2\tfrac{3}{4}$	**3** $3\tfrac{1}{5} - 1\tfrac{5}{6}$	**3** $4\tfrac{1}{12} - 3\tfrac{3}{8}$
4 $3\tfrac{1}{4} - 1\tfrac{1}{3}$	**4** $6\tfrac{1}{5} - 4\tfrac{1}{4}$	**4** $8\tfrac{1}{4} - 2\tfrac{7}{20}$
5 $3\tfrac{1}{3} - 1\tfrac{1}{2}$	**5** $6\tfrac{2}{5} - 4\tfrac{3}{4}$	**5** $5\tfrac{1}{3} - 3\tfrac{5}{6}$
6 $5 - \tfrac{1}{2}$	**6** $9 - \tfrac{3}{4}$	**6** $14 - \tfrac{1}{8}$
7 $6 - \tfrac{1}{4}$	**7** $8 - \tfrac{1}{6}$	**7** $12 - \tfrac{2}{5}$
8 $7 - \tfrac{1}{5}$	**8** $11 - \tfrac{1}{7}$	**8** $10 - \tfrac{5}{6}$
9 $10 - 6\tfrac{3}{4}$	**9** $15 - 1\tfrac{1}{2}$	**9** $161 - 21\tfrac{1}{3}$
10 $12 - 1\tfrac{1}{6}$	**10** $9 - 2\tfrac{1}{4}$	**10** $101 - 1\tfrac{5}{9}$

Multiplication of fractions

Five important rules to remember when multiplying fractions:

1. The word 'of' in a fraction sum should be changed to a multiplication sign: $\frac{1}{2}$ of $\frac{2}{3}$ should be written $\frac{1}{2} \times \frac{2}{3}$.

2. In all multiplication sums involving fractions mixed numbers must be changed to improper fractions.

3. A whole number should be made to 'look like' a fraction by writing it over 1; 17 would be written $\frac{17}{1}$.

4. The value of an answer is not changed if the numerator of one fraction and the denominator of a *different* fraction in the same sum are divided by the same common factor.

5. To obtain an answer multiply the numerators and multiply the denominators; the resulting fraction should, if necessary, be reduced to its lowest terms.

There are two main types of multiplication sum involving fractions:

A   Multiplication of fractions by fractions
B   Multiplication of fractions by mixed numbers, mixed numbers by mixed numbers, and multiplication of fractions and mixed numbers by whole numbers, in any combination

A   Multiplication of fractions by fractions

Example $\frac{2}{3} \times \frac{9}{10}$

Method   1. Write out the sum: $\frac{2}{3} \times \frac{9}{10}$

2. *Cancel* any figure on the top line of either fraction which has a common factor with *any* figure on the bottom line

You must cancel *both top and bottom lines* at the same time

In this sum the 2 and the 10 divide by 2, leaving 1 (top) and 5 (bottom)

The 9 and the 3 divide by 3 leaving 3 (top) and 1 (bottom):

$$\frac{\overset{1}{\cancel{2}}}{\underset{1}{\cancel{3}}} \times \frac{\overset{3}{\cancel{9}}}{\underset{5}{\cancel{10}}} = \frac{1 \times 3}{1 \times 5} = \frac{3}{5}$$

$\frac{3}{5}$ *Answer*

EXERCISE 24

24a
1. $\frac{1}{2} \times \frac{2}{3}$
2. $\frac{2}{5} \times \frac{1}{2}$
3. $\frac{1}{2} \times \frac{4}{5}$
4. $\frac{1}{4} \times \frac{2}{3}$
5. $\frac{3}{4} \times \frac{8}{9}$
6. $\frac{1}{5} \times \frac{10}{11}$
7. $\frac{2}{5} \times \frac{15}{16}$
8. $\frac{1}{6} \times \frac{2}{3}$
9. $\frac{5}{6} \times \frac{3}{10}$
10. $\frac{1}{7} \times \frac{7}{8}$

24b
1. $\frac{5}{9} \times \frac{3}{25}$
2. $\frac{7}{9} \times \frac{3}{14}$
3. $\frac{5}{9} \times \frac{3}{20}$
4. $\frac{4}{9} \times \frac{27}{28}$
5. $\frac{7}{10} \times \frac{20}{21}$
6. $\frac{3}{20} \times \frac{4}{15}$
7. $\frac{10}{11} \times \frac{33}{40}$
8. $\frac{5}{11} \times \frac{22}{25}$
9. $\frac{5}{12} \times \frac{24}{25}$
10. $\frac{7}{12} \times \frac{9}{14}$

24c
1. $\frac{8}{9} \times \frac{3}{4} \times \frac{1}{2}$
2. $\frac{10}{11} \times \frac{22}{25} \times \frac{1}{4}$
3. $\frac{1}{2} \times \frac{4}{5} \times \frac{10}{12}$
4. $\frac{1}{4} \times \frac{8}{9} \times \frac{27}{28}$
5. $\frac{2}{3} \times \frac{9}{22} \times \frac{11}{12}$
6. $\frac{3}{4} \times \frac{8}{15} \times \frac{5}{16}$
7. $\frac{1}{5} \times \frac{10}{11} \times \frac{22}{25}$
8. $\frac{2}{5} \times \frac{5}{6} \times \frac{3}{10}$
9. $\frac{1}{7} \times \frac{14}{15} \times \frac{5}{6}$
10. $\frac{5}{6} \times \frac{12}{13} \times \frac{39}{40}$

B Multiplication of fractions, mixed numbers, and whole numbers in any combination

(a) Multiplication of mixed numbers by fractions

Example $2\frac{1}{2} \times \frac{9}{25} \times 1\frac{1}{3}$

Method   1. Change all mixed numbers to improper fractions:

Sum now reads: $\frac{5}{2} \times \frac{9}{25} \times \frac{4}{3}$

2. *Cancel* — Divide by 5: 5 (top)
                                    25 (bottom)
             Divide by 3: 9 (top)
                                    3 (bottom)
             Divide by 2: 4 (top)
                                    2 (bottom)

$$\frac{\cancel{5}^1}{\cancel{2}_1} \times \frac{\cancel{9}^3}{\cancel{25}_5} \times \frac{\cancel{4}^2}{\cancel{3}_1} = \frac{1 \times 3 \times 2}{1 \times 5 \times 1} = \frac{6}{5}$$

3. Reduce $\frac{6}{5}$ to a mixed number: $= 1\frac{1}{5}$

$1\frac{1}{5}$ *Answer*

EXERCISE 25

25a
1. $\frac{1}{2} \times 1\frac{1}{3}$
2. $\frac{1}{3} \times 2\frac{1}{4}$
3. $\frac{1}{4} \times 1\frac{1}{3}$
4. $\frac{1}{5} \times 1\frac{2}{3}$
5. $\frac{1}{6} \times 2\frac{1}{4}$
6. $\frac{1}{7} \times 4\frac{2}{3}$

25b
1. $1\frac{1}{11} \times \frac{1}{12}$
2. $1\frac{1}{12} \times \frac{1}{13}$
3. $1\frac{1}{3} \times \frac{1}{8}$
4. $2\frac{1}{4} \times \frac{1}{12}$
5. $1\frac{1}{3} \times \frac{1}{16}$
6. $1\frac{2}{3} \times \frac{1}{20}$

25c
1. $1\frac{1}{2} \times \frac{2}{3}$
2. $1\frac{1}{2} \times 1\frac{1}{3}$
3. $1\frac{3}{4} \times 1\frac{5}{7}$
4. $2\frac{1}{5} \times 1\frac{9}{11}$
5. $2\frac{1}{3} \times 2\frac{1}{7}$
6. $1\frac{1}{6} \times 3\frac{3}{7} \times 1\frac{1}{4}$

EXERCISE 25 (*continued*)

25a	25b	25c
7 $\frac{1}{8} \times 5\frac{1}{3}$	**7** $2\frac{1}{4} \times \frac{1}{24}$	**7** $1\frac{3}{5} \times 1\frac{5}{8} \times 1\frac{4}{11}$
8 $\frac{1}{9} \times 3\frac{3}{5}$	**8** $4\frac{2}{3} \times \frac{1}{28}$	**8** $1\frac{1}{8} \times 1\frac{1}{3} \times 2\frac{1}{2}$
9 $\frac{1}{10} \times 6\frac{2}{3}$	**9** $5\frac{1}{3} \times \frac{1}{32}$	**9** $2\frac{1}{4} \times 2\frac{2}{3} \times \frac{1}{15}$
10 $\frac{1}{11} \times 7\frac{1}{3}$	**10** $3\frac{3}{5} \times \frac{1}{36}$	**10** $3\frac{5}{9} \times 1\frac{7}{8} \times \frac{3}{8}$

(b) Multiplication of a whole number by a fraction

Example $2 \times \frac{7}{12}$

Method   1. Rewrite whole number over 1: $\frac{2}{1} \times \frac{7}{12}$

2. *Cancel* — Divide by 2:   2 (top)

12 (bottom)

$$\frac{\overset{1}{\cancel{2}}}{1} \times \frac{7}{\underset{6}{\cancel{12}}} = \frac{1 \times 7}{1 \times 6} = \frac{7}{6}$$

3. Reduce $\frac{7}{6}$ to a mixed number = $1\frac{1}{6}$

$1\frac{1}{6}$ *Answer*

(c) Multiplication of a mixed number by a whole number

Example $1\frac{3}{4} \times 24$

Method   1. Change $1\frac{3}{4}$ to a mixed number and write 24 over 1:

$= \frac{7}{4} \times \frac{24}{1}$

2. *Cancel* by dividing 4 into 4 (bottom) and 24 (top):

$$= \frac{7}{1} \times \frac{6}{1} = \frac{7 \times 6}{1 \times 1} = \frac{42}{1} = 42 \quad Answer$$

(d) Use of the word 'of' for multiplication

The word 'of' is often used with

'quarter of' which should be written out as $\frac{1}{4} \times$
'half of' which should be written out as $\frac{1}{2} \times$
'three-quarters of' which should be written out as $\frac{3}{4} \times$

A further use is illustrated here:

Example $\frac{2}{3}$ of 96

Method   1. Change 'of' to × (multiply by) and write 96 over 1:

$$= \tfrac{2}{3} \times \tfrac{96}{1}$$

2. *Cancel* by dividing 3 into 3 (bottom) and 96 (top):

$$\frac{2}{\overset{}{\underset{1}{3}}} \times \frac{\overset{32}{96}}{1} = \frac{2 \times 32}{1 \times 1} = \frac{64}{1} = 64 \quad Answer$$

EXERCISE 26

26a		26b		26c	
1	$8 \times \tfrac{3}{4}$	**1**	$24 \times \tfrac{1}{3}$	**1**	$36 \times \tfrac{2}{3}$
2	$24 \times \tfrac{1}{2}$	**2**	$50 \times \tfrac{2}{5}$	**2**	$25 \times \tfrac{3}{5}$
3	$30 \times \tfrac{5}{6}$	**3**	$8 \times \tfrac{1}{4}$	**3**	$4 \times \tfrac{1}{4}$
4	$16 \times \tfrac{1}{8}$	**4**	$10 \times \tfrac{1}{5}$	**4**	$6 \times \tfrac{1}{12}$
5	$1\tfrac{1}{2} \times 12$	**5**	$1\tfrac{1}{4} \times 48$	**5**	$1\tfrac{1}{2} \times 36$
6	$1\tfrac{1}{3} \times 18$	**6**	$2\tfrac{1}{5} \times 25$	**6**	$1\tfrac{2}{7} \times 49$
7	$1\tfrac{1}{2} \times 10$	**7**	$1\tfrac{1}{6} \times 24$	**7**	$1\tfrac{1}{4} \times 8$
8	$2\tfrac{1}{4} \times 20$	**8**	$2\tfrac{1}{8} \times 16$	**8**	$2\tfrac{1}{3} \times 9$
9	$\tfrac{1}{4}$ of 32	**9**	$\tfrac{3}{4}$ of 40	**9**	$\tfrac{3}{5}$ of 30
10	$\tfrac{1}{3}$ of 9	**10**	$\tfrac{2}{3}$ of 24	**10**	$\tfrac{5}{8}$ of 40

Division of fractions

Important points to remember when dividing fractions:

Fractions do not like following a division sign. You must change the division sign ÷ to a ×, *and then turn the following fraction upside down.*

Division of fractions can involve whole numbers, mixed numbers, and/or fractions in the same sum. Remember that whole numbers should be written over 1 to make them look like a fraction and all mixed numbers must be changed into improper fractions.

Because the ÷ sign has to be changed to an × (multiply), the general rules for multiplication of fractions apply also to division of fractions.

Four examples of the division of fractions; all follow the same basic rules:

(a) Division of one fraction by another fraction

Example $\tfrac{1}{4} \div \tfrac{1}{12}$

Method   1. Change the ÷ to × and turn $\tfrac{1}{12}$ 'upside down'

Sum now reads: $\tfrac{1}{4} \times \tfrac{12}{1}$

2. *Cancel* — Divide by 4: 4 (bottom)
12 (top)

$$\frac{1}{\overset{}{\underset{1}{\cancel{4}}}} \times \frac{\overset{3}{\cancel{12}}}{1} = \frac{1 \times 3}{1 \times 1} = \frac{3}{1} = 3 \quad Answer$$

EXERCISE 27

27a

1. $\frac{1}{2} \div \frac{1}{4}$
2. $\frac{1}{3} \div \frac{1}{6}$
3. $\frac{1}{4} \div \frac{1}{8}$
4. $\frac{1}{5} \div \frac{1}{15}$
5. $\frac{1}{6} \div \frac{1}{12}$
6. $\frac{3}{4} \div \frac{3}{8}$
7. $\frac{2}{3} \div \frac{1}{6}$
8. $\frac{3}{4} \div \frac{1}{12}$
9. $\frac{2}{5} \div \frac{1}{10}$
10. $\frac{5}{6} \div \frac{1}{12}$

27b

1. $\frac{2}{3} \div \frac{4}{9}$
2. $\frac{3}{4} \div \frac{9}{16}$
3. $\frac{4}{5} \div \frac{16}{25}$
4. $\frac{2}{3} \div \frac{2}{15}$
5. $\frac{5}{6} \div \frac{5}{18}$
6. $\frac{6}{7} \div \frac{2}{21}$
7. $\frac{2}{3} \div \frac{1}{6}$
8. $\frac{3}{4} \div \frac{15}{24}$
9. $\frac{3}{5} \div \frac{1}{30}$
10. $\frac{2}{3} \div \frac{14}{15}$

27c

1. $\frac{5}{6} \div \frac{25}{36}$
2. $\frac{7}{8} \div \frac{7}{24}$
3. $\frac{2}{9} \div \frac{4}{27}$
4. $\frac{6}{7} \div \frac{3}{28}$
5. $\frac{3}{10} \div \frac{9}{100}$
6. $\frac{2}{9} \div \frac{1}{24}$
7. $\frac{11}{12} \div \frac{11}{18}$
8. $\frac{7}{8} \div \frac{7}{12}$
9. $\frac{2}{11} \div \frac{5}{22}$
10. $\frac{7}{10} \div \frac{1}{70}$

(b) Division of one mixed number by another mixed number

Example $2\frac{1}{2} \div 1\frac{1}{4}$

Method 1. Change both mixed numbers to improper fractions:

$$= \tfrac{5}{2} \div \tfrac{5}{4}$$

2. Change the ÷ to × and turn $\frac{5}{4}$ 'upside down':

$$\tfrac{5}{2} \times \tfrac{4}{5}$$

3. *Cancel* — Divide by 5: 5 (top)
5 (bottom)
Divide by 2: 2 (bottom)
4 (top)

$$= \frac{\overset{1}{\cancel{5}}}{\underset{1}{\cancel{2}}} \times \frac{\overset{2}{\cancel{4}}}{\underset{1}{\cancel{5}}} = \frac{1 \times 2}{1 \times 1} = \frac{2}{1} = 2 \quad Answer$$

(c) Division of a whole number by a fraction

Example $3 \div 1\frac{1}{2}$

Method 1. Write the whole number (3) over 1 and change the $1\frac{1}{2}$ to an improper fraction:

$$= \tfrac{3}{1} \div \tfrac{3}{2}$$

2. Change ÷ to × and turn $\frac{3}{2}$ 'upside down'

Sum now reads: $\frac{1}{1} \times \frac{2}{3}$

3. *Cancel* the 3 (top) and the 3 (bottom):

$$\frac{\cancel{3}^{1}}{1} \times \frac{2}{\cancel{3}_{1}} = \frac{1 \times 2}{1 \times 1} = \frac{2}{1} = 2 \quad Answer$$

(d) Division of a fraction by a whole number

Example $\frac{3}{4} \div 3$

Method   1. Write the whole number (3) over 1 ($\frac{3}{1}$)

Sum now reads: $\frac{3}{4} \div \frac{3}{1}$

2. Change the ÷ to × and turn $\frac{3}{1}$ 'upside down'

$= \frac{3}{4} \times \frac{1}{3}$

3. *Cancel* the 3 (top) and the 3 (bottom):

$$= \frac{\cancel{3}^{1}}{4} \times \frac{1}{\cancel{3}_{1}} = \frac{1 \times 1}{4 \times 1} = \frac{1}{4} \quad Answer$$

EXERCISE 28

28a

1 $1\frac{1}{2} \div \frac{3}{4}$
2 $1\frac{1}{3} \div \frac{2}{3}$
3 $1\frac{1}{5} \div 1\frac{1}{5}$
4 $3\frac{1}{3} \div 1\frac{2}{3}$

5 $3 \div \frac{1}{2}$
6 $3 \div \frac{3}{4}$
7 $5 \div \frac{5}{6}$

8 $\frac{1}{2} \div 5$
9 $\frac{3}{4} \div 3$
10 $\frac{10}{11} \div 5$

28b

1 $5\frac{1}{4} \div 1\frac{1}{20}$
2 $6\frac{1}{3} \div 1\frac{1}{18}$
3 $1\frac{1}{6} \div 2\frac{1}{3}$
4 $2\frac{7}{9} \div 1\frac{2}{3}$

5 $2 \div \frac{1}{9}$
6 $4 \div \frac{4}{5}$
7 $6 \div \frac{3}{7}$

8 $\frac{10}{11} \div 5$
9 $\frac{2}{3} \div 10$
10 $\frac{3}{4} \div 12$

28c

1 $4\frac{9}{10} \div 1\frac{2}{5}$
2 $1\frac{5}{6} \div 2\frac{4}{9}$
3 $6\frac{3}{10} \div 3\frac{3}{20}$
4 $2\frac{6}{7} \div 1\frac{1}{14}$

5 $4 \div 4\frac{4}{5}$
6 $11 \div 5\frac{1}{2}$
7 $3 \div \frac{6}{7}$

8 $\frac{8}{9} \div 4$
9 $1\frac{3}{4} \div 7$
10 $5\frac{1}{7} \div 12$

5 Decimal fractions

Definition

The *decimal point* separates a whole number (on the left) from a *decimal fraction* — a *part* of a whole — on the right. In the number 236.493, 236 is a whole number and .493 is a decimal fraction.

When writing decimal fractions which have no whole number it is usual to put a 0 before the point — 0.493; this makes the point more obvious and makes no difference to the value as it simply means there are no whole numbers.

The number of figures *to the right* of the decimal point is known as the *number of decimal places*; 0.493 has *three* decimal places.

Decimal fractions are *read* by naming each figure; 0.493 is read as nought point four nine three **not** nought point four hundred and ninety-three.

Decimal place values

Decimal fractions are simply a continuation of our 'whole number' system — where every figure is 10 times greater if it is moved **one** place to the left.

One unit becomes 10 (1×10) if we move it to the left and add a 0; the 0 has to be added to show the correct place.

In exactly the same way 0.1 becomes 10 times greater if we move it to the left and to the other side of the decimal point, making it 1.0 — that is, **one whole unit**.

Therefore 0.1 is one-tenth ($\frac{1}{10}$) of one unit.

If we move the figure 1 in 0.01 one place to the left it becomes $\frac{1}{10}$ (one-tenth); if we leave it as 0.01 it is $\frac{1}{10}$ of $\frac{1}{10}$ ($\frac{1}{10} \times \frac{1}{10} = \frac{1}{100}$) or one-hundredth.

Therefore, if we move the figure 1 in 0.001 one place to the left it becomes $\frac{1}{100}$; leave it where it is and it is $\frac{1}{10}$ of $\frac{1}{100}$ ($\frac{1}{10} \times \frac{1}{100} = \frac{1}{1000}$) or one-thousandth.

The values of the first three places of decimals are:

Decimal point	One-tenth ($\frac{1}{10}$) of a unit	One-hundredth ($\frac{1}{100}$) of a unit	One-thousandth ($\frac{1}{1000}$) of a unit
	4	9	3
=	Four-tenths	Nine-hundredths	Three-thousandths
=	$\frac{4}{10}$	$\frac{9}{100}$	$\frac{3}{1000}$

EXERCISE 29

29a   29b   29c

Give the value of the underlined figures in the following numbers:
(Write the answer in *figures*, e.g. 0.1 = $\frac{1}{10}$)

29a	29b	29c
1 0.1	**1** 0.213	**1** 0.85
2 0.52	**2** 0.221	**2** 0.32
3 0.167	**3** 0.243	**3** 0.232
4 0.054	**4** 0.674	**4** 0.327
5 0.295	**5** 0.418	**5** 0.641
6 0.21	**6** 0.246	**6** 0.219
7 0.12	**7** 0.41	**7** 0.065
8 0.74	**8** 0.63	**8** 0.284
9 0.96	**9** 0.208	**9** 0.529
10 0.316	**10** 0.73	**10** 0.762

Addition of decimal fractions

> Before starting an addition sum involving decimals remember that any whole numbers must be given a decimal point followed by a 0. In the example below 2 must be written as 2.0.
>
> When writing out the numbers which have to be added, make sure that all the decimal points are lined up under each other.

Example   0.2 + 2 + 2.613

Method   1. Rewrite the sum, adding .0 to the whole number:

   0.2 + 2.0 + 2.613

2. Write out the numbers, keeping the decimal points directly under each other:

0.200
2.000
+ 2.613
———
4.813

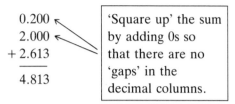

'Square up' the sum by adding 0s so that there are no 'gaps' in the decimal columns.

3. Add up all the figures as in a normal addition sum but *keep the decimal point in the right place.*

4.813   *Answer*

EXERCISE 30

30a
1 0.2 + 0.6
2 0.5 + 0.1
3 0.7 + 0.2
4 1.6 + 2.3
5 4.1 + 5.6
6 6.22 + 7.66
7 8.91 + 3.02
8 5.42 + 4
9 6.67 + 2
10 5 + 6.1 + 2.22

30b
1 0.5 + 0.6
2 0.8 + 0.3
3 0.9 + 0.4
4 1.8 + 6.4
5 7.2 + 7.9
6 5.67 + 4.33
7 9.06 + 8.07
8 6.91 + 7
9 5.96 + 11
10 11 + 12.6 + 5.62

30c
1 0.9 + 0.8
2 0.7 + 0.9
3 0.1 + 0.9
4 8.6 + 9.6
5 11.09 = 12.01
6 15.69 + 16.02
7 18.42 + 16.1
8 16.96 + 21
9 12.6 + 12 + 12.612
10 11.6 + 11 + 123

Subtraction of decimal fractions

Before starting a subtraction sum involving decimals remember that any whole numbers must be given a decimal point followed by a 0. 169 would be written 169.0.

When writing out the numbers which have to be used, make sure that all the decimal points are lined up under each other.

Note particularly:

When subtracting one number from another number the smaller number is taken from the larger number.

When decimal fractions are involved the following should be noted:

1. If both numbers contain whole numbers, the larger *whole number* is the larger number. 9.06 is larger than 8.99.

2. If only one number has a whole number, that is the larger number.

3. If both numbers are decimal fractions the following test should be done:

Example   Which is the larger decimal fraction, 0.091 or 0.81?

Method   1. Write the two numbers under each other with the points in line:

        0.091
        0.81

        2. The number containing the larger digit immediately following the point is the larger number. The two digits are 0 and 8: 8 is the larger, therefore 0.81 is the larger number.

        0.81   *Answer*

NB. Should the first two digits be equal, then comparison should be made between the two second digits following the point, and so on.

Example $126.196 - 2.3$

Method   1. Write out the sum with the larger number above the smaller number, and keeping the decimal points under each other:

$$\begin{array}{r} 126.196 \\ - 2.300 \\ \hline 123.896 \\ \hline \end{array}$$

Add 00 to 2.3 so that there are no spaces in the decimal columns

2. Proceed as with the subtraction of whole numbers, but keep the decimal point in the right place in the answer.

123.896   *Answer*

EXERCISE 31

31a
1 $4.8 - 3.2$
2 $3.6 - 1.2$
3 $10.96 - 9.85$
4 $6.44 - 2.2$
5 $3.421 - 2.31$

31b
1 $7.66 - 7.57$
2 $59.6 - 0.1$
3 $10.96 - 10.95$
4 $5.99 - 0.1$
5 $4.667 - 3.55$

31c
1 $11.1 - 0.9$
2 $66.93 - 6.84$
3 $12.361 - 11.25$
4 $6.86 - 0.2$
5 $9.552 - 8.43$

EXERCISE 31 (*continued*)

31a	31b	31c
6 3.695 − 2	**6** 2.999 − 1	**6** 2.229 − 1
7 8.08 − 6	**7** 4.99 − 3	**7** 6.95 − 6
8 7.9 − 7.65	**8** 59.6 − 59.01	**8** 6.001 − 0.001
9 19.06 − 11	**9** 22.613 − 20	**9** 100.01 − 99.01
10 8 − 6.01	**10** 9 − 0.51	**10** 6 − 5.94

Multiplication of decimal fractions

There are three main types of multiplication sums relating to decimal fractions:

A Decimal fractions multiplied by decimal fractions
B A decimal fraction multiplied by a whole number
C A whole number multiplied by a decimal fraction

A Decimal fractions multiplied by decimal fractions

(a) A straightforward sum:

Example 11.12×0.3

Method 1. Write the two numbers under each other and make a note of the decimal places in each: 11.12 (2 places)
0.3 (1 place)

Add the number of places together = 3

> This is the number of decimal places which must appear in the answer

2. Rewrite the sum *without the decimal points:* $1112 \times \cancel{0}3$
(Cross off any 0 at the beginning of a number)

3. Multiply the larger number by the smaller number:

```
    1112
  ×    3
  ──────
    3336
```

4. There must be *three decimal places* in the answer
Count back three figures from the last figure in the answer 3336 and put in the point: 3.336

3.336 *Answer*

(b) You may have to add a 0 or noughts before the figures in your answer to get the point in the right place.

Example   0.04 × 0.016

Method   1. Write the two numbers under each other and make a note of the decimal places in each: 0.04   (2 places)
0.016 (3 places)

Add the number of places together = 5

This is the number of decimal places which must appear in the answer

2. Rewrite the sum *without the decimal points:* 0̸0̸4 × 0̸0̸16
(Cross off any 0s at the beginning of the numbers)
3. Multiply the larger number by the smaller number:

16 × 4 = 64

4. There must be *five decimal places* in the answer. Add 0s *before* the 64 *until you have five places:* .00064

5. Put in the decimal point and another 0 in front of it

0.00064   *Answer*

EXERCISE 32

32a
1 0.6 × 0.2
2 0.4 × 0.3
3 0.6 × 0.4
4 0.7 × 0.2
5 0.5 × 0.3
6 0.9 × 0.2
7 0.2 × 0.8
8 0.8 × 0.4
9 0.25 × 0.05
10 3.22 × 0.01

32b
1 1.4 × 0.2
2 2.2 × 0.3
3 4.3 × 0.6
4 3.2 × 0.4
5 3.1 × 0.7
6 2.5 × 0.5
7 1.1 × 0.9
8 9.9 × 0.2
9 2.31 × 0.02
10 2.36 × 0.03

32c
1 1.21 × 0.2
2 4.32 × 0.3
3 3.45 × 0.4
4 5.41 × 0.6
5 2.11 × 0.9
6 2.1 × 0.8
7 4.3 × 0.6
8 2.6 × 0.2
9 1.21 × 0.02
10 2.23 × 0.03

B   A decimal fraction multiplied by a whole number

Example   0.605 × 9

Method   1. Count the number of decimal places in the fraction — there are *three*.

| This is the number of decimal places which must appear in the answer |

2. Rewrite the sum *without the decimal point:* 0̷605 × 9
(Cross off any 0s at the beginning of the numbers)

3. Multiply the larger number by the smaller number:

$$\begin{array}{r} 605 \\ \times \quad 9 \\ \hline 5445 \end{array}$$

4. There must be *three decimal places* in the answer.
Count back three figures from the last figure in the answer 5445 and put in the point: 5.445

5.445   *Answer*

C   A whole number multiplied by a decimal fraction:

Example   605 × 0.21

Method   1. Count the number of decimal places in the fraction — there are *two*.

| This is the number of decimal places which must appear in the answer |

2. Rewrite the sum *without the decimal point:* 605 × 0̷21
(Cross off any 0s at the beginning of the numbers)

3. Multiply the larger number by the smaller number (by long multiplication):

$$\begin{array}{r} 605 \\ \times \quad 21 \\ \hline \end{array}$$

$$\begin{array}{rl} 20 \times 605 = & 12\,100 \\ 1 \times 605 = & 605 \\ \hline 21 \times 605 = & 12\,705 \\ \hline\hline \end{array}$$

4. There must be *two decimal places* in the answer.
Count back two figures from the last figure in the answer 12 705 and put in the point: 127.05

127.05   *Answer*

EXERCISE 33

33a		33b		33c	
1	0.4 × 2	**1**	0.09 × 4	**1**	0.13 × 12
2	0.1 × 3	**2**	0.05 × 12	**2**	0.011 × 8
3	0.8 × 9	**3**	0.01 × 3	**3**	0.013 × 7
4	0.9 × 6	**4**	0.06 × 5	**4**	1.6 × 4
5	0.11 × 6	**5**	0.08 × 5	**5**	1.25 × 4
6	18 × 0.2	**6**	31 × 0.07	**6**	121 × 0.2
7	23 × 0.5	**7**	22 × 0.04	**7**	231 × 1.3
8	12 × 0.4	**8**	43 × 0.06	**8**	411 × 2.1
9	11 × 0.9	**9**	34 × 0.2	**9**	124 × 0.11
10	17 × 0.6	**10**	34 × 0.02	**10**	342 × 0.22

Multiplication of decimal fractions by multiples of 10

A   Multiplying by 10

To multiply any decimal fraction by 10, move the point one place to the right

Example   6.15 × 10

Method   1. Because 10 ends with *one* 0, move the point *one* place to the right:

6.15 × 10 = 61.5   *Answer*

To multiply any decimal fraction by a multiple of 10

Example   1.2 × 20

Method   1. Because 20 ends with *one* 0 it is a multiple of 10 — the factors of 10 are *10 and 2*

2. Multiply 1.2 × 10 by moving the point *one* place to the right = 12

3. Multiply 12 by the other factor (2): 12 × 2 = 24   *Answer*

B   Multiplying by 100

To multiply any decimal fraction by 100, move the point two places to the right

Example   3.162 × 100

Method   1.   Because 100 ends with *two* 0s, move the point *two* places to the right: 3.162 × 100 = 316.2   *Answer*

To multiply any decimal fraction by a multiple of 100

Example   2.123 × 500

Method   1. Because 500 ends with *two* 0s it is a multiple of 100 — the factors of 500 are *100 and 5*

2. Multiply 2.123 × 100 by moving the point *two* places to the right: 2.123 × 100 = 212.3

3. Multiply 212.3 by the other factor of 500 (5):

$$
\begin{array}{r}
212.3 \\
\times \quad 5 \\
\hline
1061.5 \quad \textit{Answer} \\
\hline\hline
\end{array}
$$

C   Multiplying by 1000

To multiply any decimal fraction by 1000, move the point three places to the right

Example   0.6 × 1000

Method   1. Because 1000 ends with *three* 0s, move the point *three* places to the right — to enable you to do this you must add two 0s:

0.600 × 1000 = 600 (note that you do not need the point)

600   *Answer*

To multiply any decimal fraction by a multiple of 1000

Example   4.123 × 2000

Method   1. Because 2000 ends with *three* 0s it is a multiple of 1000 — the factors of 2000 are *1000 and 2*

2. Multiply 4.123 × 1000 by moving the point *three* places to the right: 4.123 × 1000 = 4123

3. Multiply 4123 by the other factor of 2000 (2): 4123 × 2 = 8246

8246   *Answer*

EXERCISE 34

34a		34b		34c	
1	3.241 × 10	**1**	5.436 × 10	**1**	1.212 × 10
2	3.241 × 20	**2**	0.436 × 40	**2**	1.212 × 60
3	0.331 × 30	**3**	3.221 × 10	**3**	1.212 × 1000
4	0.331 × 100	**4**	3.221 × 50	**4**	1.212 × 600
5	2.133 × 100	**5**	1.143 × 100	**5**	1.212 × 6000
6	0.531 × 100	**6**	1.143 × 600	**6**	3.214 × 10
7	0.531 × 300	**7**	2.121 × 1000	**7**	3.214 × 100
8	1.432 × 1000	**8**	2.121 × 3000	**8**	3.214 × 1000
9	1.432 × 2000	**9**	3.699 × 100	**9**	3.214 × 500
10	1.432 × 3000	**10**	3.699 × 1000	**10**	3.214 × 5000

Division of decimal fractions

There are three main types of division sum relating to decimal fractions:

A Division of a decimal fraction by a decimal fraction
B Division of a decimal fraction by a whole number
C Division of a whole number by a decimal fraction

A Division of a decimal fraction by a decimal fraction

Example $0.06 \div 0.2$

Method 1. Write numbers under each other with the decimal points in line:

$$0.06$$
$$\div 0.2$$

2. Add a 0 to the shorter number so that both numbers are the same length:

$$0.06$$
$$\div 0.20$$

> It is important that both numbers have the same number of places after the decimal point — either the top or bottom must have 0s added to achieve this.

3. Cross off the decimal points and rewrite the sum: $\emptyset\emptyset6 \div \emptyset20$ (Cross off any 0s at the *beginning* of the numbers)

Sum now reads: $6 \div 20$

4. Add a decimal point and a 0 to the *first* figure: $6.0 \div 20$

5. Divide by long division method:

$$
\begin{array}{r}
0.3 \\
20\overline{)6.0}* \\
3 \times 20 = 6\,0 \\
\hline
0\,0
\end{array}
$$

* Note that as many 0s as required can be added here

0.3 *Answer*

Remember You must have a figure in the answer over every figure under the line. The decimal point must be retained in its correct position in the answer.

EXERCISE 35

35a		35b		35c	
1	1.2 ÷ 0.6	**1**	4.4 ÷ 0.11	**1**	1.24 ÷ 0.4
2	2.2 ÷ 1.1	**2**	3.6 ÷ 0.12	**2**	2.22 ÷ 0.2
3	2.6 ÷ 0.2	**3**	16.8 ÷ 0.8	**3**	2.64 ÷ 0.8
4	1.4 ÷ 0.7	**4**	6.6 ÷ 1.1	**4**	2.55 ÷ 0.5
5	2.5 ÷ 0.5	**5**	2.5 ÷ 0.5	**5**	2.48 ÷ 0.2
6	2.4 ÷ 0.6	**6**	9.9 ÷ 0.3	**6**	1.24 ÷ 0.04
7	1.5 ÷ 0.3	**7**	0.1 ÷ 0.01	**7**	2.22 ÷ 0.02
8	2.8 ÷ 0.4	**8**	4.9 ÷ 0.07	**8**	2.64 ÷ 0.03
9	1.6 ÷ 0.8	**9**	12.1 ÷ 1.1	**9**	2.88 ÷ 0.09
10	2.7 ÷ 0.3	**10**	8.1 ÷ 0.09	**10**	0.189 ÷ 0.03

B   Division of a decimal fraction by a whole number

   Example   0.36 ÷ 12

   Method   1. Write out the sum and divide by the short division method:

$$\begin{array}{r} 0.03 \\ \hline 12\overline{)0.36} \end{array}$$

   0.03   *Answer*

C   Division of a whole number by a decimal fraction

   Example   18 ÷ 0.12

   Method   1. Write the whole number followed by a decimal point and a 0:

   18.0
   ÷ 0.12

   2. Write the divisor underneath with the decimal points in line

   3. Add a 0 to the number with the fewer decimal places so that both end at the same point:

   18.00
   ÷ 0.12

It is important that both numbers have the same number of places after the decimal point — either the top or bottom must have 0s added to achieve this.

---

4. Cross off the decimal points and any 0s at the *beginning* of the numbers. Sum now reads: 1800 ÷ 012

5. Divide by short division method:

$$\frac{0150}{12)\overline{1800}}$$

150   *Answer*

EXERCISE 36

36a		36b		36c	
1	0.24 ÷ 2	**1**	6.15 ÷ 5	**1**	7.154 ÷ 7
2	2.64 ÷ 2	**2**	2.648 ÷ 2	**2**	6.018 ÷ 6
3	4.24 ÷ 4	**3**	3.693 ÷ 3	**3**	0.0105 ÷ 5
4	3.63 ÷ 3	**4**	4.124 ÷ 4	**4**	9.018 ÷ 9
5	6.66 ÷ 6	**5**	5.565 ÷ 5	**5**	2.406 ÷ 6
6	4 ÷ 0.2	**6**	50 ÷ 0.5	**6**	25 ÷ 0.05
7	5 ÷ 0.5	**7**	55 ÷ 1.1	**7**	72 ÷ 0.6
8	6 ÷ 0.3	**8**	51 ÷ 0.3	**8**	9 ÷ 0.3
9	2 ÷ 0.1	**9**	100 ÷ 0.2	**9**	64 ÷ 0.8
10	7 ÷ 0.7	**10**	156 ÷ 0.3	**10**	81 ÷ 0.09

Division of decimal fractions by multiples of 10

A   Dividing by 10

To divide any decimal fraction by 10, move the point one place to the left

Example   25.9 ÷ 10

Method   1. Because 10 ends with *one* 0, move the point *one* place to the left:
25.9 ÷ 10 = 2.59   *Answer*

To divide any decimal fraction by a multiple of 10

Example   36.6 ÷ 30

Method   1. Because 30 ends with *one* 0 it is a multiple of 10 — the factors of 30 are *10 and 3*

2. Divide 36.6 by 10 — move the point *one* place to the left = 3.66

3. Divide 3.66 by the other factor of 30 (3) = 3.66 ÷ 3

$$\frac{1.22}{3)\overline{3.66}}$$

1.22   *Answer*

B Dividing by 100

To divide any decimal fraction by 100, move the point two places to the left

Example   76.2 ÷ 100

Method   1. Because 100 ends with *two* 0s, move the point *two* places to the left:
76.2 ÷ 100 = 0.762   *Answer*
(NB.   Add a 0 before the point)

To divide any decimal fraction by a multiple of 100

Example   25.5 ÷ 500

Method   1. Because 500 ends with *two* 0s it is a multiple of 100 — the factors of 500 are *100 and 5*

2. Divide 25.5 by 100 by moving the point *two* places to the left:

25.5 ÷ 100 = 0.255   (write a 0 before the point)

3. Divide 0.255 by the other factor of 500 (5) ⟶ 5)0.255 ‾‾‾‾‾‾ (0.051)

= 0.051   *Answer*

C Dividing by 1000

To divide any decimal fraction by 1000, move the point three places to the left:

Example   36.4 ÷ 1000

Method   1. Because 1000 ends with *three* 0s, move the point *three* places to the left — to enable you to do this you must add another 0 before the 36.4;
036.4 ÷ 1000 = 0.0364   *Answer*

To divide any decimal fraction by a multiple of 1000

Example   484.4 ÷ 4000

Method   1. Because 4000 ends with *three* 0s, it is a multiple of 1000 — the factors of 4000 are *1000 and 4*

2. Divide 484.4 by 1000 by moving the point *three* places to the left:
484.4 ÷ 1000 = 0.4844

3. Divide 0.4844 by the other factor of 4000 (4) ⟶ 4)0.4844 ‾‾‾‾‾‾ (0.1211)

0.1211   *Answer*

EXERCISE 37

37a	37b	37c
1 462.6 ÷ 10	**1** 8.488 ÷ 10	**1** 3.63 ÷ 10
2 462.6 ÷ 20	**2** 8.488 ÷ 40	**2** 3.63 ÷ 100
3 142.5 ÷ 10	**3** 205.5 ÷ 10	**3** 3.63 ÷ 1000
4 142.5 ÷ 30	**4** 205.5 ÷ 50	**4** 2.48 ÷ 20
5 448.4 ÷ 100	**5** 636.6 ÷ 100	**5** 2.48 ÷ 200
6 448.4 ÷ 400	**6** 636.6 ÷ 600	**6** 2.48 ÷ 2000
7 551.5 ÷ 1000	**7** 369.3 ÷ 1000	**7** 2.48 ÷ 10
8 551.5 ÷ 5000	**8** 369.3 ÷ 3000	**8** 109.08 ÷ 100
9 621.7 ÷ 100	**9** 599.6 ÷ 100	**9** 109.08 ÷ 1000
10 621.7 ÷ 1000	**10** 599.6 ÷ 1000	**10** 109.08 ÷ 2000

Conversion of common fractions to decimal fractions

Example Express $\frac{3}{4}$ as a decimal fraction
(*or* Convert $\frac{3}{4}$ to a decimal fraction)

Method 1. Rewrite $\frac{3}{4}$ as $3 \div 4$

2. Add a decimal point and 00 to the first figure, making it 3.00 (NB. As many noughts as necessary may be added here after the point; the number will not be known until the division sum is worked out, when they can be added as required).

3. Divide, using the short division method.
Remember to put the decimal point in the answer line in the same position as it is *below* the line:

$$\frac{0.75}{4)3.00}$$

0.75 *Answer*

NB. If a mixed number has to be expressed as a decimal fraction, convert only the *fraction* to a decimal fraction; the whole number remains the same for both.

For example, $2\frac{3}{4}$ would be expressed as 2.75

EXERCISE 38

Convert the following fractions to decimal fractions:

38a	38b	38c
1 $\frac{1}{2}$	**1** $\frac{1}{10}$	**1** $\frac{17}{20}$
2 $\frac{1}{4}$	**2** $\frac{3}{10}$	**2** $\frac{19}{20}$
3 $\frac{1}{5}$	**3** $\frac{7}{10}$	**3** $1\frac{1}{2}$
4 $\frac{2}{5}$	**4** $\frac{9}{10}$	**4** $3\frac{2}{5}$
5 $\frac{3}{5}$	**5** $\frac{1}{20}$	**5** $7\frac{1}{8}$

EXERCISE 38 (*continued*)

38a	38b	38c
6 $\frac{4}{5}$	**6** $\frac{3}{20}$	**6** $8\frac{7}{8}$
7 $\frac{1}{8}$	**7** $\frac{7}{20}$	**7** $4\frac{7}{10}$
8 $\frac{3}{8}$	**8** $\frac{9}{20}$	**8** $5\frac{3}{20}$
9 $\frac{5}{8}$	**9** $\frac{11}{20}$	**9** $2\frac{11}{20}$
10 $\frac{7}{8}$	**10** $\frac{13}{20}$	**10** $6\frac{19}{20}$

Conversion of decimal fractions to common fractions

Example   Express 0.25 as a common fraction
(*or* convert 0.25 to a common fraction)

Method   1. Cross off the 0 before the decimal point ($\emptyset$.25) and write the .25 over 1; make sure that the figure 1 is written directly *under* the point:

$$\frac{.25}{1}$$

2. Write a 0 under each decimal figure and cross off the point:

The fraction now reads: $\dfrac{25}{100}$

3. *Cancel*, if required (25 will divide exactly into 25 and 100):

$$\frac{\cancel{25}^{\,1}}{\cancel{100}_{\,4}} = \frac{1}{4} \quad Answer$$

NB.   If a decimal fraction, preceded by a whole number, has to be expressed as a common fraction, convert only the decimal fraction to a common fraction — the whole number remains the same.

For example, 10.25 would be expressed as $10\frac{1}{4}$.

EXERCISE 39

39a	39b	39c
Convert the following decimal fractions to common fractions:		
1 0.5	**1** 0.05	**1** 0.08
2 0.25	**2** 0.7	**2** 0.85
3 0.75	**3** 0.45	**3** 0.16
4 0.3	**4** 0.9	**4** 0.95
5 0.4	**5** 0.55	**5** 0.36
6 0.2	**6** 0.04	**6** 0.375
7 0.8	**7** 0.15	**7** 0.125
8 0.6	**8** 0.12	**8** 0.875
9 0.35	**9** 0.65	**9** 0.625
10 0.1	**10** 0.24	**10** 5.375

6 Percentages

Percentage means the 'rate per hundred'. 5% means '5 in every hundred'.

Note that percentages can always be expressed as fractions by writing the percentage over 100. 5% is the same as $\frac{5}{100}$ — or, if the fraction is reduced to its lowest terms — $\frac{1}{20}$; this is the same as saying 1 in 20.

Percentage of a whole number

Example   25% of 120

Method   1. 25% can be written as a fraction by writing the 25 over 100: $\frac{25}{100}$

2. 'of' can be written as × (multiply by)

3. 120 can be made to 'look like' a fraction by writing it over 1

4. Rewrite sum as a fraction sum: $\frac{25}{100} \times \frac{120}{1}$

5. *Cancel* 1. Divide by 25: 25 (top)
                                    100 (bottom)

           2. Divide by 4:   4 (bottom)
                                   120 (top).

$$\frac{\overset{1}{\cancel{25}}}{\underset{4}{\cancel{100}}} \times \frac{\overset{30}{\cancel{120}}}{1} = \frac{1 \times 30}{1 \times 1} = \frac{30}{1} = 30$$

30 *Answer*

EXERCISE 40

40a
1 2% of 50
2 8% of 100
3 5% of 40
4 11% of 100

40b
1 20% of 100
2 70% of 150
3 50% of 130
4 80% of 110

40c
1 60% of 90
2 25% of 600
3 35% of 300
4 85% of 200

EXERCISE 40 (*continued*)

40a		40b		40c	
5	6% of 600	**5**	30% of 160	**5**	75% of 400
6	3% of 700	**6**	90% of 110	**6**	45% of 600
7	10% of 80	**7**	40% of 120	**7**	5% of 1000
8	9% of 200	**8**	25% of 32	**8**	10% of 1200
9	10% of 110	**9**	75% of 32	**9**	20% of 1250
10	5% of 120	**10**	15% of 20	**10**	40% of 2500

One number expressed as a percentage of another number

Example   Express 6 as a percentage of 300

Method   1. Write the smaller number over the larger number and multiply by $\frac{100}{1}$

Sum now reads $\frac{6}{300} \times \frac{100}{1}$

2. *Cancel*   1. Divide by 100: 100 (top)
                              300 (bottom)

   2. Divide by 3:   6 (top)
                      3 (bottom)

3. $\dfrac{\overset{2}{\cancel{6}}}{\underset{1}{\cancel{300}}} \times \dfrac{\overset{1}{\cancel{100}}}{1} = \dfrac{2 \times 1}{1 \times 1} = \dfrac{2}{1} = 2\%$

2%   Answer

NB.   Taking one number as a percentage of another number can result in a percentage which is a mixed number:

Example   Express 10 as a percentage of 80

Method   (as above):

$\dfrac{10}{80} \times \dfrac{100}{1}$ *Cancel* $\dfrac{\overset{1}{\cancel{10}}}{\underset{2}{\cancel{80}}} \times \dfrac{\overset{25}{\cancel{100}}}{1} = \dfrac{25}{2} = 12\frac{1}{2}\%$ *Answer*

Never leave the answer as an improper fraction

EXERCISE 41

What percentage of the second number is the first number:

41a		41b		41c	
1	5, 10	**1**	2, 8	**1**	8, 160
2	25, 50	**2**	25, 500	**2**	60, 80
3	10, 40	**3**	12, 30	**3**	150, 250

EXERCISE 41 (*continued*)

41a		41b		41c	
4	1, 100	**4**	3, 20	**4**	65, 130
5	20, 200	**5**	5, 25	**5**	6, 300
6	6, 60	**6**	15, 60	**6**	105, 210
7	40, 200	**7**	55, 110	**7**	11, 220
8	100, 400	**8**	75, 150	**8**	8, 64
9	10, 50	**9**	36, 180	**9**	75, 200
10	20, 400	**10**	100, 1000	**10**	250, 400

Conversion of common fractions to percentages

Example Express $\frac{1}{5}$ as a percentage

Method 1. Multiply the fraction by $\frac{100}{1} = \frac{1}{5} \times \frac{100}{1}$

2. *Cancel* Divide 5 (bottom) and 100 (top) by 5:

$$\frac{1}{\underset{1}{\cancel{5}}} \times \frac{\overset{20}{\cancel{100}}}{1} = \frac{1 \times 20}{1 \times 1} = \frac{20}{1} = 20\%$$

20% *Answer*

EXERCISE 42

42a

42b

42c

Convert the following common fractions to percentages:

	42a		42b		42c
1	$\frac{1}{2}$	**1**	$\frac{9}{10}$	**1**	$\frac{2}{25}$
2	$\frac{1}{4}$	**2**	$\frac{1}{20}$	**2**	$\frac{4}{25}$
3	$\frac{3}{4}$	**3**	$\frac{3}{20}$	**3**	$\frac{9}{25}$
4	$\frac{1}{5}$	**4**	$\frac{7}{20}$	**4**	$\frac{6}{25}$
5	$\frac{2}{5}$	**5**	$\frac{9}{20}$	**5**	$\frac{12}{25}$
6	$\frac{3}{5}$	**6**	$\frac{11}{20}$	**6**	$\frac{7}{25}$
7	$\frac{4}{5}$	**7**	$\frac{13}{20}$	**7**	$\frac{3}{25}$
8	$\frac{1}{10}$	**8**	$\frac{17}{20}$	**8**	$\frac{11}{25}$
9	$\frac{3}{10}$	**9**	$\frac{19}{20}$	**9**	$\frac{8}{25}$
10	$\frac{7}{10}$	**10**	$\frac{1}{25}$	**10**	$\frac{13}{25}$

Conversion of percentages to common fractions

Percentages may be expressed (a) as whole numbers, or (b) as mixed numbers.

(a) Conversion of percentages (whole numbers) to common fractions

Example Express 40% as a fraction
(*or* Convert 40% to a fraction)

Method 1. Write percentage figure → $\dfrac{40}{100}$
over 100

2. *Cancel* Divide top and bottom figures by 20 $= \dfrac{\overset{2}{\cancel{40}}}{\underset{5}{\cancel{100}}} = \dfrac{2}{5}$

$\frac{2}{5}$ *Answer*

(b) Conversion of percentages (mixed numbers) to common fractions:

Example Express $33\frac{1}{3}$% as a fraction
(*or* Convert $33\frac{1}{3}$% to a fraction)

Method 1. Express percentage figure as a top-heavy (improper) fraction: $33\frac{1}{3} = \frac{100}{3}$

2. Multiply *denominator* by 100 $= \dfrac{100}{3 \times 100}$

3. *Cancel* 100 (top) and 100 (bottom) $= \dfrac{\overset{1}{\cancel{100}}}{\underset{1}{3 \times \cancel{100}}} = \dfrac{1}{3}$

$\frac{1}{3}$ *Answer*

EXERCISE 43

Convert the following percentages to common fractions:

43a		43b		43c	
1	1%	**1**	30%	**1**	50%
2	2%	**2**	7%	**2**	60%
3	3%	**3**	8%	**3**	80%
4	4%	**4**	9%	**4**	11%
5	5%	**5**	15%	**5**	13%
6	10%	**6**	35%	**6**	17%
7	20%	**7**	40%	**7**	19%
8	25%	**8**	45%	**8**	85%
9	$2\frac{1}{2}$%	**9**	$12\frac{1}{2}$%	**9**	$62\frac{1}{2}$%
10	$7\frac{1}{2}$%	**10**	$37\frac{1}{2}$%	**10**	$87\frac{1}{2}$%

Conversion of decimal fractions to percentages

In **all** cases the *method* is to multiply the decimal fraction by 100 by moving the decimal point **two** places to the right. Remaining decimal fractions should be expressed as common fractions.

Note the following examples:

One figure only after the decimal point:

Express 0.3 as a percentage

$= 0.30 \times 100 = 30\%$ *Answer*

Two figures after the decimal point, the first being a 0:

Express 0.03 as a percentage

$= 0.03 \times 100 = 3\%$ *Answer*

Two figures after the decimal point:

Express 0.33 as a percentage

$= 0.33 \times 100 = 33\%$ *Answer*

Three figures after the decimal point, the first two being 00s:

Express 0.005 as a percentage

$= 0.005 \times 100 = 0.5$ (Change 0.5 to a common fraction)

$= \frac{1}{2}\%$ *Answer*

Three figures after the decimal point, the first being a 0:

Express 0.025 as a percentage

$= 0.025 \times 100 = 2.5$ (Change .5 to a common fraction)

$= 2\frac{1}{2}\%$ *Answer*

Three figures after the decimal point:

Express 0.375 as a percentage

$= 0.375 \times 100 = 37.5$

(Change .5 to a common fraction)

$= 37\frac{1}{2}\%$ *Answer*

EXERCISE 44

44a	44b	44c

Express the following decimal fractions as percentages:

44a		44b		44c	
1	0.1	**1**	0.2	**1**	0.7
2	0.5	**2**	0.6	**2**	0.8
3	0.05	**3**	0.04	**3**	0.03
4	0.08	**4**	0.06	**4**	0.07
5	0.57	**5**	0.72	**5**	0.52
6	0.59	**6**	0.96	**6**	0.66
7	0.025	**7**	0.075	**7**	0.065
8	0.005	**8**	0.035	**8**	0.045
9	0.175	**9**	0.275	**9**	0.425
10	0.225	**10**	0.325	**10**	0.675

Conversion of percentages to decimal fractions

Percentages may consist of (a) one figure only, (b) two figures, or (c) a mixed number

(a) Conversion of a percentage containing **one figure only**, to a decimal fraction

Example Express 2% as a decimal fraction

Method 1. Write down the percentage number with a decimal point: 2.0

2. Divide 2.0 by 100 (Move the decimal point two places to the left

= .02 *Answer*

(b) Conversion of a percentage containing **two figures** to a decimal fraction

Example Express 25% as a decimal fraction

Method 1. Write down the percentage number with a decimal point: 25.0

2. Divide 25.0 by 100 (Move the decimal point two places to the left) = 0.25 *Answer*

(c) Conversion of a percentage which is a **mixed number** to a decimal fraction

(i) Where the whole number is a single figure:

Example Express $2\frac{1}{2}\%$ as a decimal fraction

Method 1. Write down the percentage number as a decimal: 2.5

2. Divide 2.5 by 100 (Move the decimal point two places to the left) = 0.025 *Answer*

(ii) Where the whole number contains two figures:

Example Express $12\frac{1}{2}\%$ as a decimal fraction

Method 1. Write down the percentage number as a decimal: 12.5

2. Divide 12.5 by 100 (Move the decimal point two places to the left) = 0.125

(iii) Where the percentage is expressed as a fraction only:

Example Express $\frac{1}{2}\%$ as a decimal fraction

Method 1. *Note that $\frac{1}{2}\%$ means $\frac{1}{2}$ in every 100*

The ratio $\frac{1}{2}:100 = 5:1000$ (a percentage of $\frac{5}{1000}$)

$\frac{5}{1000}$ expressed as a decimal fraction = 0.005

$\frac{1}{2}\% = 0.005$ *Answer*

From this can be worked out:

$\frac{1}{4}\% = 0.005 \div 2 = 0.0025$ ($\frac{1}{2}\% \div 2$)
$\frac{3}{4}\% = 0.0025 \times 3 = 0.0075$ ($\frac{1}{4}\% \times 3$)

EXERCISE 45

45a

Express the following percentages as decimal fractions:

	45a		45b		45c
1	1%	1	5%	1	6%
2	4%	2	3%	2	2%
3	15%	3	35%	3	45%
4	75%	4	97%	4	65%
5	50%	5	30%	5	40%
6	$2\frac{1}{2}\%$	6	$4\frac{1}{2}\%$	6	$8\frac{1}{2}\%$
7	$12\frac{1}{2}\%$	7	$27\frac{1}{2}\%$	7	$62\frac{1}{2}\%$
8	$17\frac{1}{2}\%$	8	$32\frac{1}{2}\%$	8	$47\frac{1}{2}\%$
9	$22\frac{1}{2}\%$	9	$37\frac{1}{2}\%$	9	$87\frac{1}{2}\%$
10	$\frac{1}{2}\%$	10	$\frac{1}{4}\%$	10	$\frac{3}{4}\%$

7 Ratio

Ratio is a way of expressing a comparison between one number and another number. It is made up of numbers with the sign : between them; 4:5 is read as '4 is to 5' or 'the ratio is 4 to 5'.

Ratios should be expressed in the simplest terms possible. The ratio 5:10 is the same as the ratio 1:2 — each number in 5:10 can be divided by 5 to produce the simplest expression.

To reduce a ratio to its simplest terms

Example Express 56:84 as simply as possible

Method 1. Write the given ratios on one line with a space between them:

56 : 84

2. Divide through *all* numbers by any factors common to all:

$$\begin{array}{r} 4) \underline{56 : 84} \\ 7) \underline{14 : 21} \\ 2 : 3 \end{array}$$

Continue dividing until there are no further common factors

3. Ratio in its simplest terms = 2:3

2:3 *Answer*

EXERCISE 46

46a

Express the following ratios in their simplest terms:

46a	46b	46c
1 10:20	**1** 4:10	**1** 8:14:28
2 5:25	**2** 6:9	**2** 21:24:27
3 3:9	**3** 20:24	**3** 24:28:32
4 6:12	**4** 15:25	**4** 25:45:90
5 4:12	**5** 4:28	**5** 48:54:72

EXERCISE 46 (*continued*)

46a		46b		46c	
6	3:15	**6**	15:40	**6**	28:84:98
7	2:14	**7**	24:30	**7**	56:72:80
8	5:20	**8**	21:49	**8**	45:63:81
9	3:24	**9**	40:64	**9**	40:80:120
10	2:20	**10**	18:81	**10**	72:144:288

8 Proportion

A *proportion* is a share of something in relation to another relative share, or shares, of the same thing.

Quantities can be divided into any number of *proportional parts*.

Example   Divide 660 in the proportion 3:2:1

Method   1. Add up the proportions: (3:2:1) = 3 + 2 + 1 = 6

2. Divide the given quantity by 6: $\longrightarrow$ $6\overline{)660}$ as $\dfrac{110}{}$

3. List the proportions and multiply each by 110:

3 × 110 = 330
2 × 110 = 220
1 × 110 = 110
——
660 ✓

| Add these numbers together; if the answer agrees with the given quantity the sum is correct. |

The proportional parts are 330, 220 and 110
(Make sure you list them in the same order as shown in the question)

330, 220, 110   *Answer*

EXERCISE 47

47a
1   Divide 6 in the ratio 2:1
2   Divide 15 in the ratio 4:1
3   Divide 12 in the ratio 4:2
4   Divide 16 in the ratio 6:2
5   Divide 24 in the ratio 5:1
6   Divide 18 in the ratio 5:4
7   Divide 14 in the ratio 4:3
8   Divide 32 in the ratio 5:3
9   Divide 30 in the ratio 3:2
10 Divide 21 in the ratio 5:2

47b
1   Divide 22 in the ratio 5:6
2   Divide 27 in the ratio 2:7
3   Divide 33 in the ratio 3:8
4   Divide 66 in the ratio 2:4
5   Divide 70 in the ratio 3:4
6   Divide 100 in the ratio 3:7
7   Divide 60 in the ratio 1:2:3
8   Divide 77 in the ratio 1:2:4
9   Divide 27 in the ratio 2:3:4
10 Divide 24 in the ratio 3:4:5

47c
1   Divide 600 in the ratio 3:2:1
2   Divide 1800 in the ratio 4:3:2
3   Divide 936 in the ratio 6:2:1
4   Divide 1100 in the ratio 6:4:1
5   Divide 630 in the ratio 6:2:1
6   Divide 450 in the ratio 4:3:2
7   Divide 660 in the ratio 3:2:1
8   Divide 396 in the ratio 5:4
9   Divide 330 in the ratio 8:2:1
10 Divide 720 in the ratio 9:2:1

9 Average (mean)

The *average*, or *mean*, of a group of numbers, is the sum of all the numbers divided by the *number of numbers* in the group.

Example Find the average (*or* mean) of 128, 40, 152 and 144

Method 1. *Add* the quantities together:
$$
\begin{array}{r}
128 \\
40 \\
152 \\
+\,144 \\
\hline
\text{TOTAL} = 464
\end{array}
$$

2. Divide the answer by the *number* of items in the addition

$$
\begin{array}{r}
116 \\
4\overline{)464}
\end{array}
$$

3. The average (or mean) is 116

116 *Answer*

EXERCISE 48

Find the average of each of the following sets of numbers:

48a		48b		48c	
1	16, 18	**1**	13, 8, 15	**1**	3, 21, 24, 20
2	26, 40	**2**	27, 12, 24	**2**	29, 48, 63, 60
3	24, 42	**3**	46, 66, 29	**3**	44, 45, 15, 72
4	14, 44	**4**	30, 33, 21	**4**	51, 6, 30, 33
5	32, 46	**5**	42, 6, 45	**5**	57, 69, 12, 66
6	30, 48	**6**	9, 51, 36	**6**	9, 36, 39, 56
7	20, 50	**7**	4, 58, 28	**7**	4, 64, 28, 36
8	36, 52	**8**	36, 8, 52	**8**	48, 8, 68, 32
9	22, 60	**9**	24, 60, 33	**9**	52, 24, 60, 72
10	38, 58	**10**	48, 12, 48	**10**	44, 84, 12, 80

10 **Median**

The *median* is the figure in the middle position of a series of values.

To find the median of an odd number of values

Example   Find the median of 5, 4, 5, 6, 8, 3, 7

Method   1. Write the numbers out in numerical order: 3, 4, 5, ⑤ 6, 7, 8

         2. Ring round the *middle* figure (5); this is the **median**

         5   *Answer*

To find the median of an even number of values

Example   Find the median of 2, 4, 9, 6

Method   1. Write the numbers out in numerical order: 2, 4, 6, 9

         2. Ring round the **two** middle numbers and add them together:

         2, ④, ⑥ 9

         $4 + 6 = 10$

         3. Divide the answer by 2: $10 \div 2 = 5$

         5 is the **median**

         5   *Answer*

EXERCISE 49

49a   49b   49c

Find the median of the following sets of numbers:

	49a		49b		49c
1	4, 3, 7	**1**	3, 9, 4, 5, 2	**1**	5, 4, 5, 6, 8, 3, 7
2	1, 6, 2	**2**	1, 4, 5, 8, 3	**2**	3, 6, 5, 7, 8, 4, 1
3	7, 3, 2	**3**	6, 9, 2, 7, 4	**3**	6, 14, 11, 10, 9, 5, 4
4	3, 1, 5	**4**	1, 5, 3, 6, 2	**4**	1, 7, 9, 11, 13, 5, 3
5	8, 4, 9	**5**	7, 3, 2, 8, 6	**5**	1, 13, 14, 9, 12, 1, 2

EXERCISE 49 (*continued*)

49a		49b		49c	
6	4, 1, 5, 2	**6**	1, 3, 6, 7, 9, 4	**6**	1, 2, 8, 10, 6, 4, 12, 6
7	9, 1, 7, 10	**7**	2, 1, 7, 5, 8, 3	**7**	5, 2, 9, 3, 7, 1, 8, 10
8	5, 3, 2, 7	**8**	5, 2, 6, 1, 7, 3	**8**	8, 1, 8, 4, 6, 3, 14, 2
9	6, 9, 2, 4	**9**	8, 1, 4, 9, 1, 2	**9**	3, 1, 4, 15, 2, 9, 8, 10
10	6, 8, 4, 1	**10**	3, 8, 4, 9, 7, 5	**10**	5, 3, 6, 16, 8, 4, 9, 11

11 Mode (or modal number)

This is the number which occurs most often in a series of values.

Example Find the mode of 8, 5, 6, 9, 11, 6, 5, 7, 6

Method 1. Write down the series in the order given:

8, 5, 6, 9, 11, 6, 5, 7, 6

2. Write down the first number (8) and put a tick by it for every 8 you can *cross off* in your list.
(It is important to cross off each number as it is dealt with)

8√

Next, write down   5√√   (Cross off both 5s)

6√√√

Proceed until all numbers have been dealt with	9√
	11√
	7√

3. 6 occurs more often than any other number; 6 is the **mode**

6   *Answer*

EXERCISE 50

Find the mode of the following sets of numbers:

50a		50b		50c	
1	2, 1, 1	**1**	12, 2, 12, 22, 32	**1**	1, 2, 3, 1, 2, 3, 1
2	4, 2, 4	**2**	12, 22, 32, 22, 42	**2**	10, 8, 3, 9, 4, 5, 10, 4, 7, 10
3	9, 1, 9	**3**	6, 56, 16, 26, 56	**3**	18, 20, 20, 19, 20, 18, 20, 19
4	5, 6, 6	**4**	31, 13, 1, 13, 3	**4**	4, 4, 2, 5, 1, 5, 5, 1, 3, 2
5	8, 3, 8	**5**	29, 19, 9, 29, 39	**5**	3, 4, 5, 6, 3, 4, 5, 6, 3, 4, 5, 6, 3
6	1, 16, 16, 5	**6**	11, 12, 11, 12, 11, 13	**6**	1, 2, 2, 1, 3, 3, 1, 4, 4, 1, 5, 5
7	17, 4, 19, 4	**7**	16, 6, 1, 61, 16, 16	**7**	29, 39, 49, 49, 29, 39, 49
8	12, 2, 22, 2	**8**	19, 91, 19, 19, 91, 9	**8**	0.4, 0.5, 0.6, 0.7, 0.6, 0.4, 0.5, 0.4
9	19, 5, 91, 5	**9**	72, 27, 2, 27, 22, 7	**9**	33, 333, 3, 3, 3, 33, 333, 33, 333, 33
10	60, 3, 30, 3	**10**	51, 15, 51, 15, 51, 5	**10**	$\frac{3}{4}, \frac{1}{2}, \frac{2}{3}, \frac{1}{2}, \frac{3}{4}, \frac{2}{3}, \frac{1}{2}$

12 British money

General rules

In all calculations involving money it is important to:

(a) Change **all** amounts of money into £'s

e.g.   3p should be written as £0.03
          15p should be written as £0.15

(b) **All whole** £'s should be written with a point and 2 0s

e.g.   £5 should be written as £5.00

Addition of money

Example   £1.20 + 5p + 20p + £10.54

Method   1. Check on (a) and (b) above

Sum now reads: £1.20 + £0.05 + £0.20 + £10.54

2. Write out sum with the points under each other

```
      £
     1.20
     0.05
     0.20
 +  10.54
   ------
   £11.99
```

Add up columns as in the addition of decimal fractions, starting with the column on the right, and retaining the point in the correct position

£11.99   *Answer*

EXERCISE 51

51a
1   20p + 59p
2   67p + 23p
3   66p + 53p
4   £1.06 + 59p

51b
1   27p + 72p
2   87p + 3p
3   76p + 42p
4   £3.21 + 68p

51c
1   59p + 30p
2   57p + 33p
3   65p + 44p
4   £5.11 + 27p

EXERCISE 51 (*continued*)

51a
5 £2.16 + 84p
6 £4.19 + 95p
7 £5.92 + 8p
8 £5.92 + £6 + 10p
9 £12 + £6.92 + £10
10 12p + £51 + 3p

51b
5 £16.51 + 49p
6 £7.64 + 59p
7 £4.96 + 4p
8 £11 + £12.36 + 3p
9 £19 + £13 + 2p
10 96p + 4p + £15

51c
5 £56.10 + 90p
6 £54.67 + 48p
7 £69.91 + 9p
8 £54.19 + £6 + 6p
9 £56 + £61 + 1p
10 96p + 99p + £96

Subtraction of money

Before starting your calculations refer to the general rules for calculations involving money on page 74.

Example £5 − 7p

Method 1. After referring to rules, rewrite sum: £5.00 − £0.07

2. Write out sum with the points under each other:

£
5.00
− 0.07
———
£4.93
═══

Proceed as with subtraction of decimal fractions, retaining point in the correct position

£4.93 *Answer*

EXERCISE 52

52a
1 56p − 42p
2 99p − 78p
3 58p − 8p
4 £2.98 − £1.56
5 £3.16 − £3.09
6 £5.55 − £4.61
7 £6.50 − 50p
8 £3.59 − £2.70
9 £6 − 36p
10 £16 − 4p

52b
1 79p − 53p
2 52p − 12p
3 37p − 15p
4 £3.86 − £1.42
5 £5.61 − £5.59
6 £3.67 − £2.76
7 £11.40 − 40p
8 £6.56 − £5.61
9 £12 − 59p
10 £12 − 9p

52c
1 89p − 56p
2 67p − 57p
3 36p − 5p
4 £6.92 − £6.91
5 £7.66 − £4.67
6 £12.16 − £11.21
7 £12.75 − 75p
8 £17.39 − £10.42
9 £56 − 90p
10 £3 − 51p

Multiplication of money

Before starting your calculations refer to the general rules for calculations involving money on page 74.

Example   98p × 3

Method   1. After referring to rules, rewrite sum: £0.98 × 3:

$$
= \begin{array}{r}
£ \\
0.98 \\
\times \quad 3 \\
\hline
£2.94 \\
\hline
\end{array}
$$

> Proceed as for multiplication of decimal fractions by whole numbers, retaining the point in its correct position

£2.94   *Answer*

EXERCISE 53

53a
1   9p × 2
2   18p × 4
3   16p × 5
4   £1.11 × 3
5   £2.21 × 6
6   12p × 9
7   13p × 8
8   20p × 7
9   £1.20 × 6
10   £5.60 × 10

53b
1   8p × 4
2   15p × 3
3   50p × 2
4   £1.20 × 8
5   £5.12 × 6
6   25p × 5
7   12p × 10
8   40p × 7
9   £10.01 × 11
10   £9.21 × 9

53c
1   7p × 3
2   12p × 2
3   19p × 5
4   £1.60 × 4
5   £60.11 × 7
6   30p × 6
7   10p × 11
8   12p × 9
9   £11.11 × 8
10   £10.01 × 10

Division of money

Before starting your calculations refer to the general rules for calculations involving money on page 74.

Example   £2.45 ÷ 7

Method   1. Write out the sum:

$$
\begin{array}{r}
£0.35 \\
\hline
7)£2.45
\end{array}
$$

Proceed as for division of decimal fractions by whole numbers, retaining the point in the correct position

£0.35   *Answer*

EXERCISE 54

54a		54b		54c	
1	18p ÷ 3	**1**	27p ÷ 9	**1**	84p ÷ 4
2	96p ÷ 6	**2**	98p ÷ 2	**2**	95p ÷ 5
3	49p ÷ 7	**3**	56p ÷ 8	**3**	77p ÷ 11
4	£2.52 ÷ 2	**4**	£9.66 ÷ 6	**4**	£8.47 ÷ 7
5	£7.50 ÷ 5	**5**	£7.50 ÷ 3	**5**	£10.89 ÷ 9
6	£9.44 ÷ 8	**6**	£12.10 ÷ 11	**6**	£56.60 ÷ 4
7	£9.90 ÷ 11	**7**	£1.17 ÷ 9	**7**	£6.40 ÷ 8
8	£7.70 ÷ 7	**8**	£10.10 ÷ 10	**8**	£78.30 ÷ 6
9	£15 ÷ 5	**9**	£64 ÷ 4	**9**	£36 ÷ 12
10	£22.11 ÷ 11	**10**	£36.36 ÷ 12	**10**	£10.30 ÷ 10

Percentages of sums of money

These calculations can be worked out in exactly the same way as the percentage of a whole number — *if the sum involves £s only*.

Example   20% of £180

Method   1. 20% of £180 $= \frac{20}{100} \times \frac{180}{1}$

Cancel by 20, then by 5: $\frac{\overset{1}{\cancel{20}}}{\underset{\underset{1}{\cancel{5}}}{\cancel{100}}} \times \frac{\overset{36}{\cancel{180}}}{1} = \frac{36}{1} = £36$

£36   *Answer*

If there are £s and pence involved in the calculation it is easier to use the following methods:

Example A   Find 10% of £561.20

Method   1. Write down the amount and underline it: £561.20

2. Write down the amount **one place to the right** (this = 10%):   £56.12   *Answer*

Example B   Find 1% of £561.20

Method   1. Write down the amount and underline it: £561.20

2. Write down the amount **two places to the right** (this = 1%):   £5.61 *   *Answer*

*to the nearest penny

78

Example C   Find 11% of £561.20

Method   1. Write down the amount and underline it: £561.20

10% of £561.20 (see Example A above)   =   56.12

 1% of £561.20 (see Example B above)   =   5.61

11% of £561.20 (10% + 1% = 11%)   =   £61.73   *Answer*

This method can be used for most calculations.

Note also the following short methods for finding percentages of sums of money which are made up of pounds and pence:

% of sum	Method
2%	Find 1% and multiply by 2
3%	Find 1% and multiply by 3
4%	Find 1% and multiply by 4
5%	Find 10% and divide by 2
9%	Find 10% and subtract 1%
15%	Find 10% and 5% and add together
19%	Find 20% and subtract 1%
20%	Divide the original sum by 5
25%	Divide the original sum by 4
30%	Find 10% and multiply by 3
40%	Find 10% and multiply by 4
50%	Divide the original sum by 2

EXERCISE 55

55a
1 10% of £1.50
2 20% of £1.50
3 10% of £6.60
4   5% of £6.60
5 15% of £6.60
6 10% of £20
7   1% of £20
8 11% of £20
9 10% of £50
10 50% of £50

55b
1 10% of £9.60
2 20% of £9.60
3 10% of £25
4   5% of £25
5 15% of £25
6 10% of £16.50
7 20% of £16.50
8 30% of £16.50
9 10% of £17.50
10 20% of £17.50

55c
1 10% of £15.40
2 20% of £15.40
3 10% of £50
4   5% of £50
5 15% of £50
6   1% of £120.10 ⎤ to
7   2% of £120.10 ⎬ nearest
8   4% of £120.10 ⎦ penny
9 10% of £267.50
10 60% of £267.50

Index

Notes

Notes

Notes

Notes